SpringerBriefs in Environmental Science

SpringerBriefs in Environmental Science present concise summaries of cutting-edge research and practical applications across a wide spectrum of environmental fields, with fast turnaround time to publication. Featuring compact volumes of 50 to 125 pages, the series covers a range of content from professional to academic. Monographs of new material are considered for the SpringerBriefs in Environmental Science series.

Typical topics might include: a timely report of state-of-the-art analytical techniques, a bridge between new research results, as published in journal articles and a contextual literature review, a snapshot of a hot or emerging topic, an in-depth case study or technical example, a presentation of core concepts that students must understand in order to make independent contributions, best practices or protocols to be followed, a series of short case studies/debates highlighting a specific angle.

SpringerBriefs in Environmental Science allow authors to present their ideas and readers to absorb them with minimal time investment. Both solicited and unsolicited manuscripts are considered for publication.

More information about this series at http://www.springer.com/series/8868

Veronika Zuzulová • Jaroslav Vido
Bernard Šiška

Agricultural Drought in Slovakia: An Impact Assessment

NDVI and Satellite Based Data

 Springer

Veronika Zuzulová
Department of Environmental Management,
Faculty of European Studies and Regional
Development
Slovak University of Agriculture in Nitra
Nitra, Slovakia

Jaroslav Vido
Department of Natural Environment,
Faculty of Forestry
Technical University in Zvolen
Zvolen, Slovakia

Bernard Šiška
Department of Environmental Management,
Faculty of European Studies and Regional
Development
Slovak University of Agriculture in Nitra
Nitra, Slovakia

ISSN 2191-5547   ISSN 2191-5555 (electronic)
SpringerBriefs in Environmental Science
ISBN 978-3-030-42060-4   ISBN 978-3-030-42061-1   (eBook)
https://doi.org/10.1007/978-3-030-42061-1

This Springer imprint is published by the registered company Springer Nature Switzerland AG.
The registered company address is: Gewerbestrasse 11, 6330 Cham, Switzerland

Preface

You are reading a publication that summarizes the possibilities of practical application of selected remote-sensing techniques in Slovakia in relation to monitoring drought and its impacts in agricultural land.

Drought affects agricultural activity in a very fundamental way, and its timely and accurate identification can reduce both economic and environmental impacts.

Because of this, we bring you the publication titled *Agricultural Drought in Slovakia: An Impact Assessment NDVI and Satellite Based Data*, in which you will read all the aspects you need to know to apply these advanced techniques to your farm. The book will also find readers among students of agricultural and forestry sciences.

We wish you a pleasant reading.

Nitra, Slovakia   Veronika Zuzulová
Zvolen, Slovakia   Jaroslav Vido
Nitra, Slovakia   Bernard Šiška

Introductory Quotation

The book deals with evaluation of agricultural drought in Slovakia. The aim is an assessment of Normalized Difference Vegetation Index (NDVI) as a method suitable for an evaluation of drought in agricultural land. Dry seasons in the time series from 1960 to 2014 were determined according to the monthly Palmer Drought Severity Index (PDSI). For this purpose, 12 sites based on limit climatic parameters were chosen. The results showed the alternation of dry and wet seasons during the period. The significant seasons of drought common for all sites were found during the years 1964, 1968, 1973, 1974, 1990, 1993, 2003, 2007, and 2012. The analysis of linear trends showed that the arid trend was identified on most sites, especially on the south part of Slovakia. The monthly NDVI at district level derived from satellite imagery at 250 m spatial resolution was applied for the period from 2000 to 2014. The correlation of results of both methods during the main growing season was determined over a common 15 years. The results showed that NDVI is an appropriate index for evaluation of drought. The impact of drought on crop yields in each month was determined by the correlation between the average monthly NDVI for each district and district yields of selected crops. For this evaluation, spring barley, winter wheat, and maize were chosen. The sensitivity of spring barley to drought influence was noticed particularly in production areas of western Slovakia, on Podunajská nížina (Danube) Valley in May. The influence of drought on winter wheat yields was found mainly in April almost in all regions of Slovakia, except for Prešov. The sensitivity of maize to drought influence was observed in July especially in Trnava and Nitra regions. The book was supported by VEGA research projects funded by the Science Grant Agency of the Ministry of Education, Science, Research and Sport of the Slovak Republic (No. 1/0370/18: Assessing the vulnerability of selected natural and disturbed ecosystems to hydrometeorological extremes, and No. 1/0767/17: Response of ecosystem services of grape-growing country to climate change regional impact – change of functions to adaptation potential) and project funded by the Slovak Research and Development Agency (APVV-18-0347).

Contents

Chapter 1
Introduction/Motivation

Abstract Chapter deals with short introduction about motivation of the presented book. The motivation is based on the need to reduce drought impacts on agricultural production in conditions of changing climate. From our point of view, is necessary to build information base and monitoring competence not only on the state level, but on the end user site. That premise led us to write the presented book. To bring the possibilities and explain advantages of using high-end methods to end users and their daily routine. The book is also applicable as practical tool for students of agriculture or forestry sciences. Although the book is based on Slovak conditions, we believe, that content could be useful and applicable for farmers from other countries.

Keywords Climate change · Agriculture · Drought · Food production · Technology

Climate change has a serious impact on agriculture as well as on other economic sectors (e.g. forestry, energetics, agro-tourism, economy). Agro-climatic analyses showed that the weather is one of the limiting factors in agricultural production and its effect could be stronger in the future.

Changes in atmospheric composition affect global air temperature, spatial and temporal distribution of precipitation, evaporation, runoff, snow cover and soil moisture. Variation of meteorological elements results in an increased frequency of ecological disturbance, for instance, drought, severe thunderstorm, forest fires, insect and disease outbreaks. The disturbances can have a negative impact on the human food supply (Backlund et al. 2008).

Drought is generally a natural phenomenon, manifested in the absence of atmospheric precipitation usually in combination with high evapotranspiration during a period, which may be several days, weeks or months. It affects all areas of life, as water is an essential condition. Therefore, drought can be considered a natural disaster.

V. Zuzulová et al., *Agricultural Drought in Slovakia: An Impact Assessment*, SpringerBriefs in Environmental Science, https://doi.org/10.1007/978-3-030-42061-1_1

Technological development has enabled to invent methods of assessment the Earth's surface features. One such method is also remote sensing.

Vegetation indices are used as a tool to evaluate vegetation cover through remote sensing. Vegetation indices are algorithms that use mathematical combinations of the spectral reflectance of vegetation, most often in the visible and near-infrared part of the spectrum (Viña et al. 2011). On the basis of collected data from satellite systems vegetation indices can be used together with selected methods of statistical analysis e.g. to estimate average yields of cultivated field crops per spatial unit (Nováková et al. 2010). The use of data from satellite imagery in various areas (e.g. meteorology, agriculture, forestry, and water management) is nowadays current.

The aim of the study was to find out the possibility of using the Normalized Differenced Vegetation Index derived from MOD13Q1 (MODIS) in the assessment of drought in the agricultural land of Slovakia. We selected three crops from cereals, which represent a major part of field crop production. Spring barley (*Hordeum vulgare* L.), which is the most widespread spring cereal in Slovakia. During the growing season, it uses the sum of spring rainfall. Winter wheat (*Triticum aestivum* L.), which represents the most significant share of cereals and is an essential source of human nutrition. During the growing season, it uses the water supplies of the winter period. Maize (*Zea mays* L.), which mainly uses the sum of summer rainfall during the growing season.

Chapter 2
State of the Art

Abstract There is no doubt that Climate change significantly affects climate dynamics of the agricultural land. Higher dynamics of the atmospheric conditions leads to higher losses in agricultural production and other secondary losses due to unpredictability of the weather. The first step to success agro production is acceptance of the agroclimatic regionalization. Although this seems as matter of course, there are many examples of bad practice. Therefore, the chapter starts with this point. In the chapter also dynamics, spatial and vertical aspects of the climatic parameters are explained. In the second part of the chapter, examples of drought impact and their potential in future climate are examined. In the end, an overview of drought assessment methods often used in agricultural landscape is presented.

Keywords Agroclimatic regionalization · Drought climatology · Agriculture · Drought impact · Drought assessment

2.1 Agroclimatic Regionalization

Regional conditions of Slovak republic are influenced first of all by altitudinal profile of Slovakia. Temperature and water balance conditions are given in Table 2.1.

Actual agro-climatic regionalization is evaluated according to meteorological data from years 1961 to 1990. Spatial distribution is given on Fig. 2.1. Classification of macro regions and regions are the same as in work Kurpelová et al. (1975). This classification is useful also from the point of view of defining of traditional agricultural production zones. Warm macro region cover conditions of maize and sugar beet agro regions, moderately warm macro region fit to potato agro region and cold macro region to mountainous production agro region.

V. Zuzulová et al., *Agricultural Drought in Slovakia: An Impact Assessment*, SpringerBriefs in Environmental Science, https://doi.org/10.1007/978-3-030-42061-1_2

Table 2.1 Agro climatic regions of Slovak Republic

	Macro region	Region	*TS10*	*Eo-R*
1	Cold	–	<2000	<0
2	Moderately warm	–	2000–2400	0–50
3	Warm	Wet	2400–2600	50–100
4		Wet – normal	2600–2800	100–150
5		Dominantly dry	2800–3000	150–200
6		Dry	>3000	>200

TS 10 sum of daily mean temperature >10 °C, *Eo* potential evapotranspiration in mm, *R* precipitation in mm

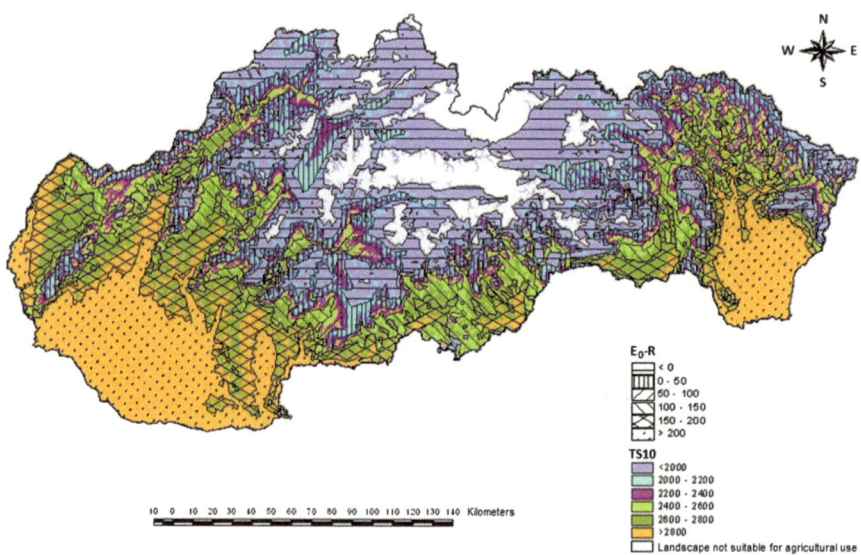

Fig. 2.1 Agroclimatic regionalization of Slovak Republic

Some details on present climate are given in next part, where comparison in context of climate change is given.

Climatic stations used for GIS analyses in this report were selected both from the point of view of altitude (up to 900 m a.s.l. – upper border of plant production) and spatial distribution. Evaluated acreage represents 45,000 km² – 90% of total area of Slovakia Selected stations represent four agro regions as given in Table 2.2.

Table 2.2 Agricultural zones and related climatic stations

Agricultural regions (productive type)	Altitude m a.s.l.	Climatic stations (name)	Altitude m a.s.l.
Maize	<200	Somotor	100
		Hurbanovo	115
		Nitra	143
		Piešťany	165
		Kamenica n/C	178
Sugar beet	200–350	Rimavská Sobota	214
		Prievidza	260
		Košice	230
		Sliač	330
Potato	300–650	Bardejov	304
		Sliač	330
		Liptovský Hrádok	640
Mountainous	> 600	Liptovský Hrádok	640

2.2   Agriculture and Climate Change Impacts in Slovakia

2.2.1   Air Temperature and Potential Evapotranspiration

Ongoing climate change causes a constant increase of air temperature and hence potential resp. real evapotranspiration. This is mainly related to the intensification of the greenhouse effect associated with increased greenhouse gas emissions. The fifth report of the International Panel for Climate Change (IPCC 2014) highlighted that only the least likely scenario of RCP 2.6 climate change (radiation gain of 2.6 W.m^2) is expected to keep the global temperature rise below 2 °C compared to the pre-industrial period. The other scenarios (RCP 4.5, RCP 6.0, RCP 8.5) do not assume that the air temperature increase will be below 2 °C until 2100 (Fig. 2.2). This situation is particularly unfavourable since RCP 4.5 assumes a global temperature rise of 2–2.5 °C and RCP 6.0 up to 2–3 °C by 2050–2100. In this regard, we remain optimistic and hope that global efforts to mitigate climate change will not break. Otherwise, there is a catastrophic variant of the RCP 8.5 scenario, which assumes a global air temperature increase of 2–4.5 ° C in the horizon of 2050–2100 (Fig. 2.2).

Such an unprecedented rapid increase of air temperature will entail a number of negative phenomena, such as: the disruption of the thermohaline cycle and sea ice in the northern hemisphere, the rapid changes in temporal and spatial distribution of precipitation, or the change in atmospheric circulation conditions.

Because of current geopolitical situation and increasing trend of fossil fuels consumption by emerging countries, the plan to keep the temperature rise below 2 °C does not seem feasible. This is also supported by the withdrawal of the U.S. from the Paris Conference in June 2017. Therefore, it is very rational approach to be prepared for the development of the RCP 4.5 and RCP 6.0 scenarios.

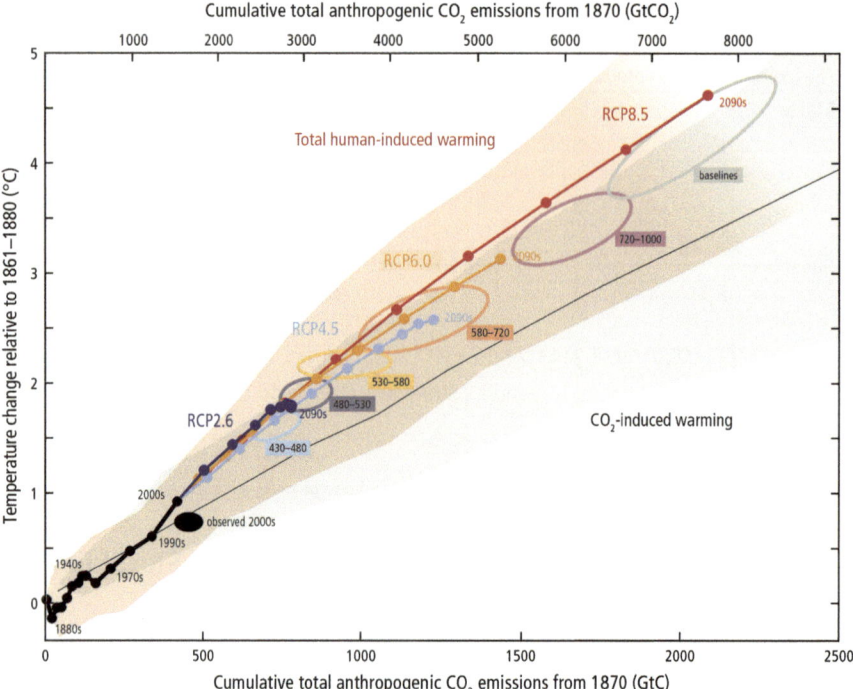

Fig. 2.2 RCP scenarios and anticipated air temperature (https://archive.ipcc.ch/report/graphics)

Regional downscaling of the future (anticipated) climate is summarized in Škvarenina et al. (2018). Results showed that according to RCP 2.6 scenario, we can expect a gradual increase in air temperature of 1.6 °C in 2050 compared to the reference period 1986–2005.

By middle of the twenty-first century, the average air temperature should not increase as of the RCP 2.6. According to the RCP 4.5 scenario, the air temperature will be increasing until 2060, with the mean deviation of the air temperature stabilizing at approximately +2.4 °C compared to the reference period 1986–2005.

The situation is more critical in the RCP 6.0 scenario. This scenario anticipate that the temperature will continue to rise steadily until the end of the century, with an estimated deviation from normal (1986–2005) of 3 °C around 2100. The most unfavourable scenario RCP 8.5 represents an extreme temperature increase for the territory of Slovakia at a level of more than 3.5 °C compared to the 1986–2005 normal.

The upward trend of air temperature expected in the next decades is also confirmed by measurements on various meteorological stations in Slovakia since the second half of the twentieth century.

Detailed analysis based on meteorological data from the Arboretum Mlyňany station showed significant increase of air temperature by 1.5 °C between 1966 and 2013 (Fig. 2.3) (Vido et al. 2019). This confirmed general upward trend of air temperature that is projected for next decades.

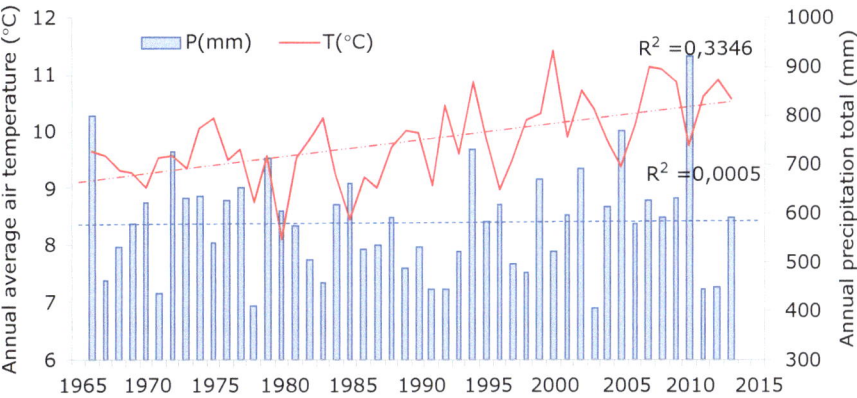

Fig. 2.3 Trend analysis of average annual air temperature and annual precipitation totals at the station in Arboretum Mlyňany in 1966–2013

Vilček et al. (2016) analyzed dynamics of air temperature development at the stations Michalovce, Hurbanovo, Rožňava, Sliač, Oravská Lesná and Skalnaté Pleso. Study confirmed significant increase of air temperature on all the investigated stations. Results also showed that air temperature increase also in the cold part of year (Fig. 2.4). This increase resulted also in early onset of spring phenological phases of forest trees in Slovakia described by Zverko et al. 2014. The earlier onset of phenological phases and higher air temperatures in the cold half-year result to higher potential evapotranspiration. Higher potential evapotranspiration is demonstrated by Hrvoľ et al. (2009) and Střelcová et al. (2006).

2.2.2   Precipitation

The most comprehensive study of precipitation characteristics in Slovakia was published in Zeleňáková et al. (2017). In order to describe changes in the precipitation trends over a studied region, annual, seasonal, and monthly precipitation distributions from both temporal and spatial perspectives were described by authors.

The annual mean precipitation total in Slovakia during the period 1981–2013 was 720.2 mm. Highest mean monthly precipitation amounts were recorded in June (87.6 mm), followed by July (86.2 mm) and May (78.9 mm). On the other hand, the lowest precipitation amounts were in February and January, with 39.8 mm and 44.1 mm, respectively. More precisely, this annual distribution actually had two maxima: one in June, as described above, and a secondary one in November with an average of 54.8 mm of precipitation (Fig. 2.5).

The absolute highest precipitation total on an annual scale was in 2010 when 2075 mm were recorded at the Jasná meteorological station in the Nízke Tatry Mountains (unit 7 in Fig. 2.6). In contrast, the lowest amount was recorded in 2011

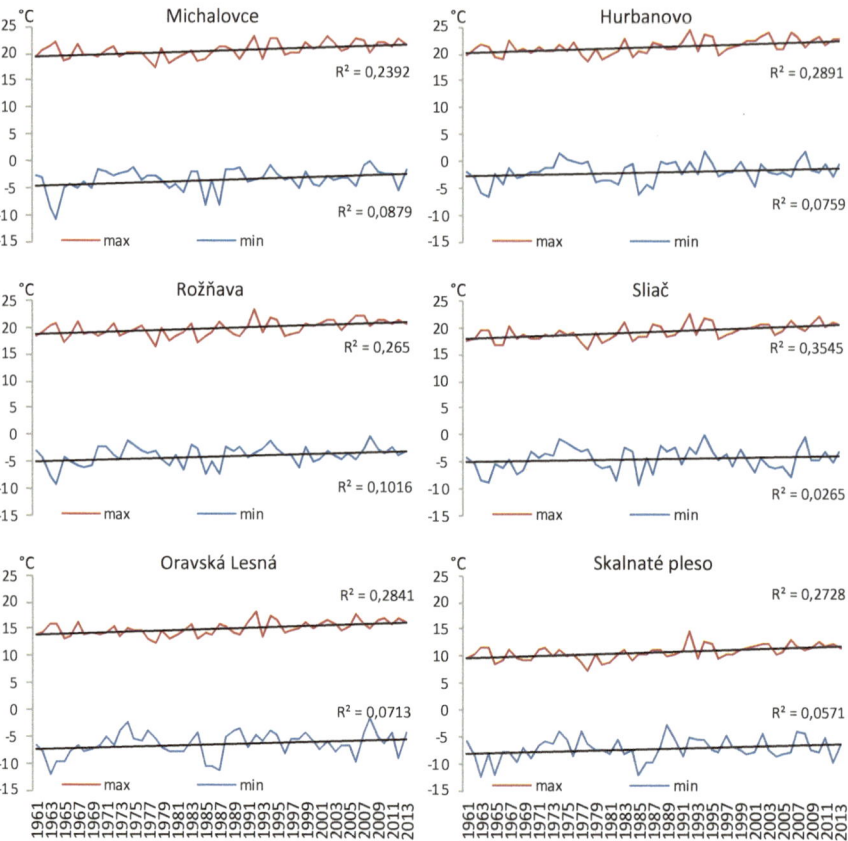

Fig. 2.4 Linear trends of monthly mean temperatures of the coldest (Tmin) and warmest month (Tmax) at selected meteorological stations (Vilček et al. 2016)

at the Malé Kosihy meteorological station located in the Podunajská nížina (Danube) Valley (unit 4 in Fig. 2.6).

2.2.2.1   Spatial Distribution of Annual Mean Precipitation Distribution

The highest totals were recorded in the mountainous areas located in north and northwestern Slovakia (Western Carpathian Mountains) and in northeastern Slovakia (Eastern Carpathian Mountains). On the other hand, the driest areas were located in the lowland (nížina) areas extending from the west to the southeast of the country: Borská nížina, Podunajská nížina, Juhoslovenská nížina, and Východoslovenská nížina. These lowlands form parts of the Pannonian Basin (Carpathian Basin) geomorphological unit at its northernmost borders. However, low precipitation totals were also recorded on the leeward of the Western Carpathian Mountains: Podtatranská kotlina Basin, Hornádska kotlina Basin, Šarišská

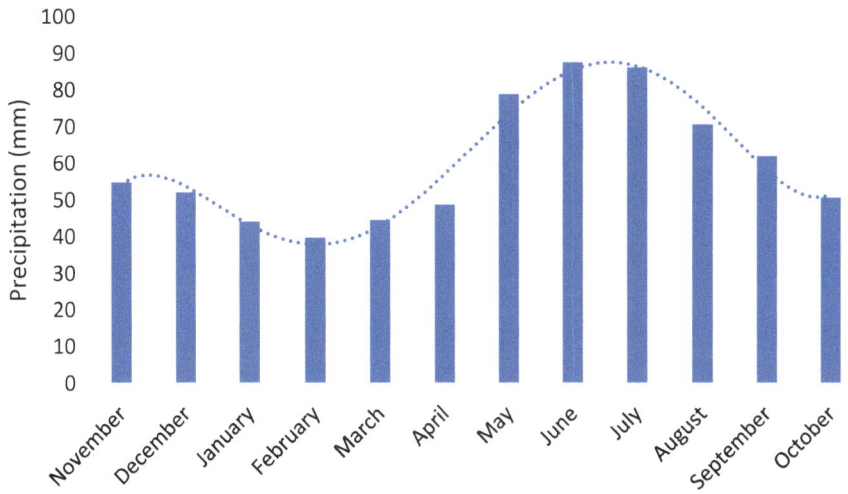

Fig. 2.5 Mean monthly precipitation totals over Slovakia within the period 1981–2013. The wave is depicted by sixth order polynomial trend (Zeleňáková et al. 2017)

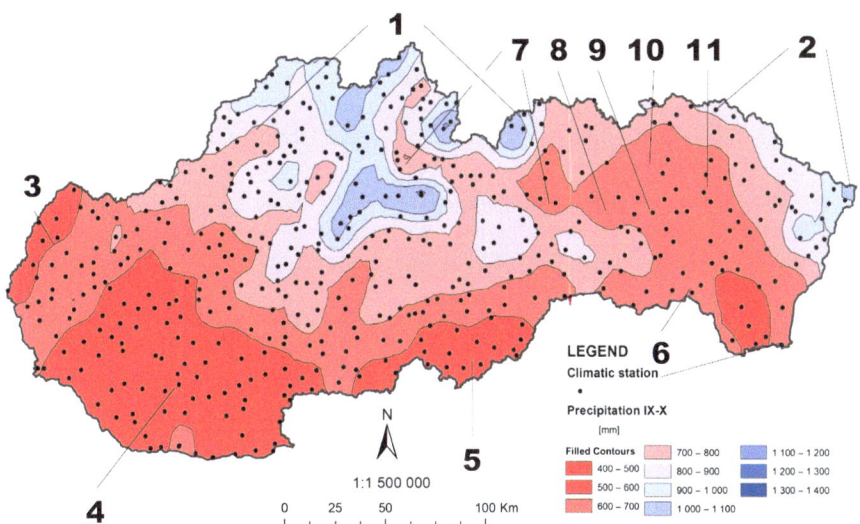

Fig. 2.6 Annual precipitation distribution over Slovakia in the period 1981–2013. Numbers in the figure represent geomorphological units: *1*. Western Carpathian Mts., *2*. Eastern Carpathian Mts., *3*. Borská nížina Valley, *4*. Podunajská nížina Valley, *5*. Juhoslovenská nížina Valley, *6*. Východoslovenská nížina Valley, *7*. Podtatranská kotlina Basin, *8*. Hornádska kotlina Basin, *9*. Šarišská vrchovina Highlands, *10*. Čergov Mts., *11*. Ondavská vrchovina Highlands (Zeleňáková et al. 2017)

vrchovina Highlands, Čergov Mountains, and Ondavská vrchovina Highlands. Annual precipitation totals are spatially depicted in Fig. 2.6.

2.2.2.2   Spatial Precipitation Distribution on a Monthly Scale

The highest rain amounts were recorded from May to July, especially in the mountainous regions of the Western and Eastern Carpathians (Fig. 2.7). In contrast to this, the rain shadow of the Western Carpathians was evident in the area of the Podtatranská kotlina Basin as well as in the Hornádska kotlina Basin, especially in July. In addition, these areas, as well as the Šarišská vrchovina Highlands, Čergov Mountains, and Ondavská vrchovina Highlands were the driest parts of the country in the period from December to February, drier than the southern parts of the country that are generally the driest areas during the whole year. This region was very dry also in October. As suggested above, areas with generally low precipitation amounts in all months are lowlands elongated from the southwest to the southeast of the country. Although these areas are the most important agricultural regions of the country, very low precipitation amounts were recorded here even in the rainiest season from April to July. The driest month in this area was January (Zeleňáková et al. 2017).

2.2.2.3   Annual, Seasonal, and Monthly Precipitation Trends

In general terms, the study of the 33 years of records in 487 Slovakian rain gauges revealed an increase in the summer precipitation and, conversely, a decrease in the autumn precipitation. Table 2.3 summarizes the results from the trend analysis based on the Mann–Kendall nonparametric statistical test.

The table shows that for the annual precipitation (based on annual precipitation totals), 157 of the rain gauges (32%) revealed significant trends. Of those stations, 155 showed an increasing trend and only two (located in the north of Slovakia) showed a decreasing trend. The magnitude of those trends varied between −0.61 mm/year at the Oravská Polhora station (situated in northern Slovakia) and 0.92 mm/year at the Vyšná Boca station (situated in the middle of the country), where the highest positive trend in precipitation was detected. Although the general precipitation trend showed a significant increase, the annual trends presented a very slight increase.

Winter and spring precipitation showed more increasing (in 68 + 20 rain gauges) than decreasing (in 3 + 1 rain gauges) trends, although the percentage of significant trends was 15% in winter and only 4% in spring. The MK test found significant trends in 25% of the stations in summer, all of them indicating increasing precipitation. In autumn, only one rain gauge denoted a significant trend, also in the same sense (Table 2.3).

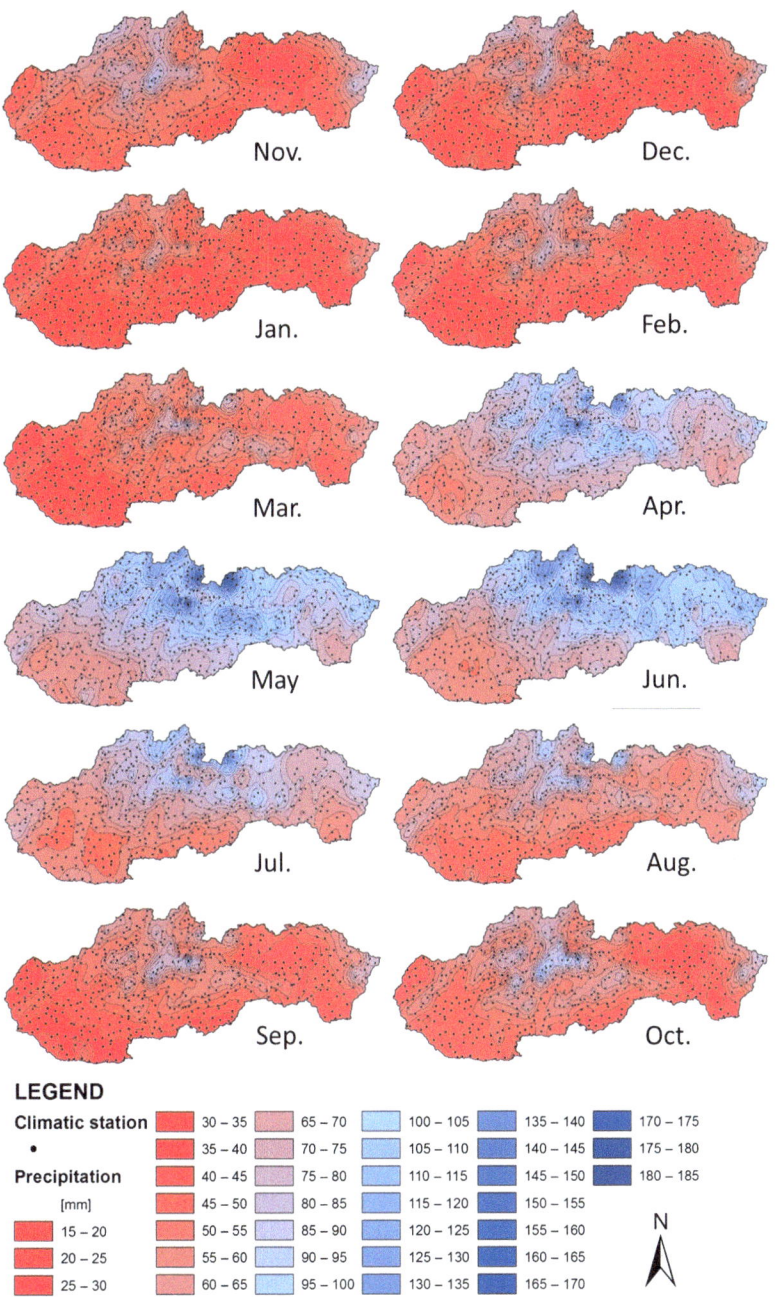

Fig. 2.7 Monthly precipitation distribution over Slovakia in the period 1981–2013 (Zeleňáková et al. 2017)

Table 2.3 Summary statistics of the precipitation trends in the annual, seasonal, and monthly data series for 487 stations in Slovakia during the period 1981–2013 (Zeleňáková et al. 2017)

Precipitation	Number of significant decreasing trends	Number of significant increasing trends	% of significant trends	Number of stations with no trends
Annual	2	155	32	330
Winter	3	68	15	416
Spring	1	20	4	466
Summer	0	123	25	364
Autumn	0	1	0	486
November	0	1	0	486
December	15	36	10	436
January	0	105	22	382
February	0	26	5	461
March	0	13	3	474
April	3	0	1	484
May	2	0	0	485
June	1	48	10	438
July	0	245	50	242
August	3	1	1	483
September	0	0	0	487
October	0	2	0	485

The Mann–Kendall trend test was also applied to detect temporal and spatial trends in the seasonal and monthly precipitation time series. Figure 2.8 presents the spatial distribution of monthly precipitation trends in Slovakia between 1981 and 2013, using the Theil–Sen Estimator value for each of the 487 gauge stations. It is evident that, from the monthly point of view, strong changes in precipitation distribution were recorded in the investigated period 1981–2013.

Regarding the monthly precipitation data, the analysis showed that the highest percentage of significant increasing trends occurred in July: 245 rain gauges (i.e., 50% of the 487 analyzed) showed significant positive trends. Also in summer, namely in June, 10% of rain gauge stations denoted significant increasing trends. This also happened for the winter season, where 22% and 7% of the rain gauges denoted significant increasing trends in January and December, respectively. However, in December increasing trends occurred mainly in the south and southeast of Slovakia, while 3% of significant decreasing trends were revealed in December mainly in the north of Slovakia. This is very interesting, considering the presence of winter precipitation in the form of snow especially in the mountainous areas of central and northern Slovakia. Therefore, decreasing precipitation trends in December in mountainous areas could imply a potential problem in the spring hydrological situation because the replenishment of ground water is dependent on snow cover. There was a high number of increasing trends in precipitation in January and February, as well as in June and mostly July. Although the percentage of significant trends in January was the second highest (22%) after July (50%), it is evident that the trend slope is

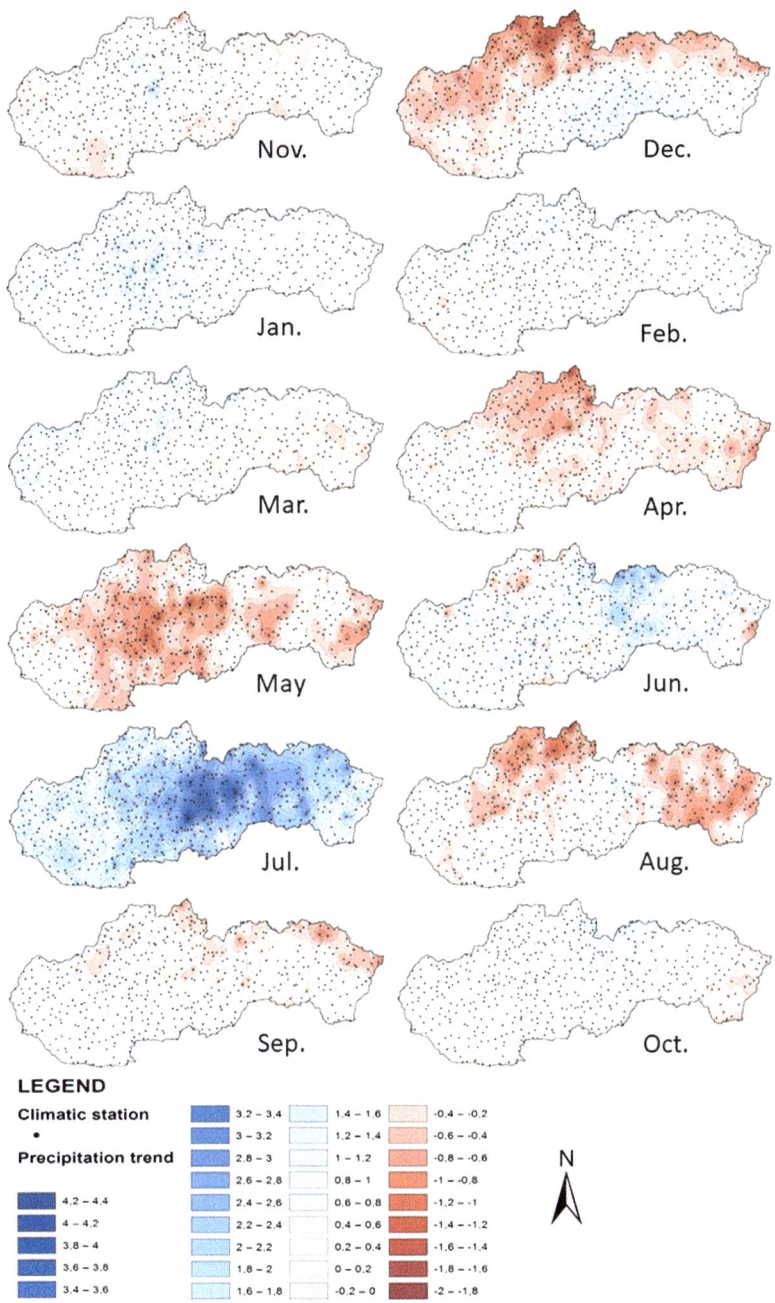

Fig. 2.8 Spatial distribution of monthly precipitation trends (mm/year) in Slovakia during 1981–2013 (Zeleňáková et al. 2017)

relatively low (Fig. 2.8). Arid trends were recorded in April, May, August, and, partially in the northern parts of Slovakia, during December, as mentioned above. The pronounced spatial variability of precipitation indicates prolonged drought episodes in the critical spring season (agricultural drought), and, on the other hand, frequent flood events occurrence in mountainous areas mainly in the summer season.

2.2.3   Drought Climatology

Regarding to information gained in previous chapters it is evident that drought climatology of Slovakia depends on various parameters e.g. altitudinal levels, orientation to prevailing winds and season of year (Vido et al. 2019).

Škvarenina et al. (2009a) pointed out that altitude and topography are strong climate-differentiating factors. Consequently, under conditions of considerably variable topography in the West Carpathians, drought plays an extra-important role. The primary importance of climate from the point of view of the natural vegetation has already been pointed out by prof. Zlatník in 1976. The author defines the vegetation stages as basic units characterizing altitudinal climate conditions (vertical differentiation) through vegetation (biocenoses). The Slovak territory has been divided by prof. Zlatník into altitudinal vegetation stages named after the significant tree or bush indicator species dominating the area. These stages are characterized by their dominant climax tree species as follows: stage 1, oak (*Quercus*) vegetation, stage 2, beech-oak (*Fagus-Quercus*) vegetation, stage 3, oak-beech (Quercus-Fagus) vegetation, stage 4, beech (*Fagus*) vegetation, stage 5, fir-beech (Abies-Fagus) vegetation, stage 6, spruce-fir-beech (*Picea-Abies-Fagus*) vegetation, stage 7, spruce (*Picea*) vegetation, stage 8, mountain pine (*Mughetum*) vegetation, stage 9, alpine vegetation (non-forest high mountain pastures). The vegetation stages of lower elevations, i.e., oak vegetation (stage 1), oak vegetation with admixture of beech (stage 2), and beech vegetation with admixture of oak (stage 3) are rather arid during the vegetation period (from March to September). The precipitation deficit reaches 100–300 mm during the vegetation season. The beech vegetation (stage 4) is characterized by an equitable climatic water balance. The climate humidity increases in higher vegetation stages (beech vegetation with fir (stage 5) and fir with beech and spruce (stage 6). Humidity belongs to the fundamental properties of mountain forests. The water balance reaches the highest values in the eighth vegetation stage of mountain dwarf pine and the ninth alpine stage, where the amount of precipitation considerably exceeds the evaporation requirements of the atmosphere. Within the annual balance, the surplus of precipitation water is approximately 1000 mm. Altitudinal analysis of drought climatology was in study of Škvarenina et al. (2009a, b) based on method of relative evapotranspiration which is a suitable measure of the water sufficiency for vegetation. Based on this methods, typical meteorological stations representing specific altitudinal levels were selected (Hurbanovo – 1. stage – 115 m a.s.l., Myjava – 2. stage – 375 m a.s.l., Kamenica nad Cirochou – 3. stage – 178 m a.s.l., Plaveč – 4. stage – 488 m a.s.l., Červený Kláštor – 5.

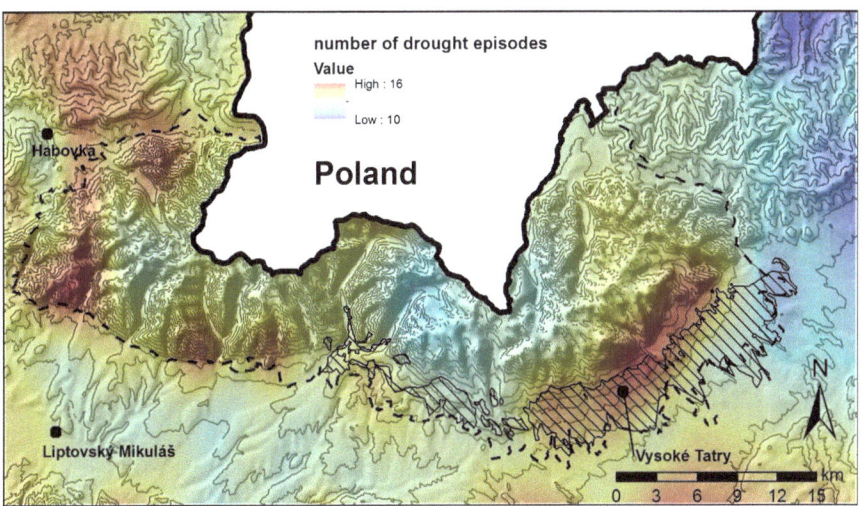

Fig. 2.9 Number of all drought situations according to the 12-month SPI. Red color represents higher frequency of drought situations. Black points represent settlements, dashed line represents border of the Tatra National Park and hatched area represents area of the windstorm of 2014. (Vido et al. 2015)

stage – 474 m a.s.l., Oravská Lesná – 6. stage – 780 m a.s.l., Ždiar – 7. stage – 1020 m a.s.l., Štrbské Pleso – 8. stage – 1360 m a.s.l.).

The results presented in Škvarenina et al. (2009a, b) point out the significant trend to aridity at the station in Hurbanovo, situated in the southern part of Slovakia. This station represents the regions of most intensive agricultural production in Slovakia. Frequent occurrence of drought as well as its rising frequency negatively influences the production of the main crops in this region as well as in other regions of Central Europe (Hlavinka et al. 2009; Dubrovsky et al. 2009). Furthermore, a statistically significant tendency to more intensive dry episodes in the region were stated by recent studies (Brázdil et al. 2009). The next two stations with elevations below 500 m a.s.l. show relatively good water supplies as Eo/P drops below 1. The stations situated to the north of the climatic line (Štrbské Pleso, Ždiar-Javorina, and Červený Kláštor) show an significant humid trend during 1951–2007.

However also orientation of the particular locality to prevailing winds shape the potential of drought occurrence. This pattern is very specific especially in mountainous regions of Central and North Slovakia. Vido et al. (2015) showed that drought occurrence is influenced by the precipitation shadow of the Tatra Mts. range and surrounding mountains situated north and to northwest of the Tatra Mts. Thus, the occurrence of drought is more likely at the south and south-east regions of the mountains than at the north/northeast windward part of the Tatra Mountains. In addition, another drought prone area was also indicated in the West Tatra Mts. This area is influenced by the Oravské Beskydy and Oravská Magura Mts. located to the northwest (Fig. 2.9).

2.2.3.1 Drought trends in Slovakia

In previous chapter some geographical aspects of drought occurrence have been
pointed out. Although this factor is important for local drought climatology, it is
important to point out trends of drought episodes. Drought trends in Slovakia have
been analyzed in several studies (Vido et al. 2015, 2019; Fendeková et al. 2018;
Zuzulová and Šiška 2017; Hrvoľ et al. 2009; Škvarenina et al. 2009a, b). General
feature of the above-mentioned studies is that drought become a serious phenomenon affecting ecosystem services. Vido et al. (2019) imply that drought trend
between the years of 1961 and 2013 tend to worse climatic conditions (Fig. 2.10).

In addition, authors analyzed the reasons. Since drought analyses of the above-
mentioned paper rely on SPEI (Standardized Precipitation Evapotranspiration
Index) method they found that the negative drought trend depends on increasing air
temperature rather than on precipitation (with no trend in analyzed period). This
imply that increasing temperature in future, as described in climate change section
will bring more frequent drought episodes in the region. From the agricultural
drought impact perspective is also very important to know what drought trends
(negative or positive) are recorded in individual months of year. Therefore, the study
of Vido et al. (2019) analyzed also trend of SPEI by individual months in the period
of 1966–2013 in the area of high agricultural importance (Podunajská nížina Valley).
The results showed that, during the analyzed period, the first 3 months of the year
showed an increasing trend toward a more humid climate; although this trend was
not significant in January and February, it was significant only at the $\alpha = 0.25$ significance level in March. A significant difference and trend were observed in April,
when we confirmed a significant increase ($\alpha = 0.05$) of the number of dry periods,
i.e., arid trend. In May, we found a nonsignificant trend indicating the improved
climate balance in relation to drought. All summer months (June, July, and August)
showed arid trends with statistical significance between $\alpha = 0.25$ and $\alpha = 0.1$. In
September and October, we did not record any changes in the occurrence frequency
of dry or humid episodes. In November, we found decreasing (arid), statistically

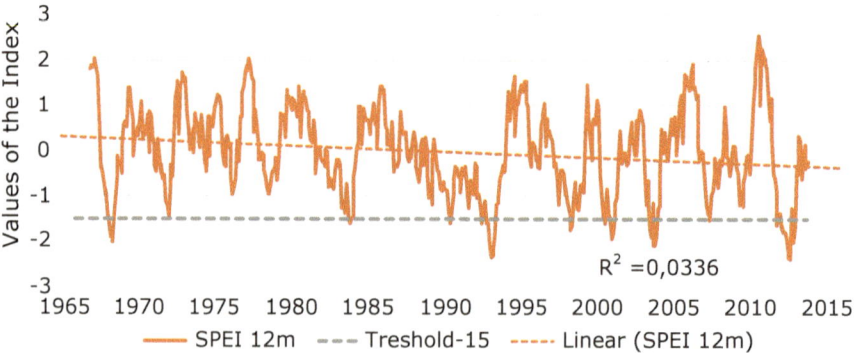

Fig. 2.10 Temporal course of a 12-month SPEI with the linear trend of the time series and an
indication of the threshold value of the index (-1.5) (Vido et al. 2019)

little significant trend ($\alpha = 0.25$). The end of the year corresponded with the projected climate change scenarios for the region in winter months (Lapin et al. 2010), i.e., humid trend, but was statistically insignificant. The trends of the months in the period between April and August corresponded with the results of Lapin et al. (2010), who analyzed the development of air moisture deficit or the increase of evapotranspiration in the region. The significant arid trend in April was the main warning signal indicating the adverse projection of the risk of drought in spring months under both optimistic and pessimistic climate scenarios in the future. Since April is one of the most important months from the agricultural view-point (Potopová et al. 2015a, b) and because of spring sowing and restoration of crops after winter, we consider this information very important.

2.2.4 Impact of Drought on Agriculture

2.2.4.1 Impact on Potential Crop Yields

Impacts of drought on agriculture in Slovakia has been previously studied mainly on Podunajská nížina Valley and Východoslovenská nížina Valley and surrounding (Vido et al. 2019; Zuzulová and Šiška 2017; Labudová et al. 2017; Takáč 2013). Results of the mentioned studies imply significant impact of drought on agriculture production in Slovakia. However, it depends on the season of the drought occurrence. For instance, oilseed rape yield deviation from normal depend on sufficient (no drought) agrometeorological conditions from March to August and especially in April (Vido et al. 2019), but for instance Winter Barley or Winter Wheat depend also on sufficient conditions in Autumn and Winter seasons (Labudová et al. 2017). In addition, economic impacts on agriculture depend also on local economic structure. Therefore, in regions with higher economic power are impacts lower and vice versa (Vido et al. 2019). Trend analyses of the August value of the SPEI for 6 months (what represent critical time period for oilseed rape yield) showed that in the period of 1961–2013 is evident negative trend (Fig. 2.11). This imply negative outlooks for agricultural production in next decades due to climate change.

2.2.4.2 Drought and Growing Season in Changed Climate Conditions

Comprehensive study of climate change impact on growing season in Slovakia was published by Šiška and Takáč (2009). Authors imply that onset, end, and duration of the vegetation period in the territory of Slovakia are limited by climatic conditions from great lowlands and mountains in Slovakia. Lowlands are usually represented by the climate station Hurbanovo on Podunajská nížina Valley or Somotor on Východoslovenská nížina Valley respectively. The locations with the highest altitude are usually represented by the climate station Telgárt for agricultural purposes. The onset of GVP (Great Vegetation Period) is accelerating, ending is delaying, and

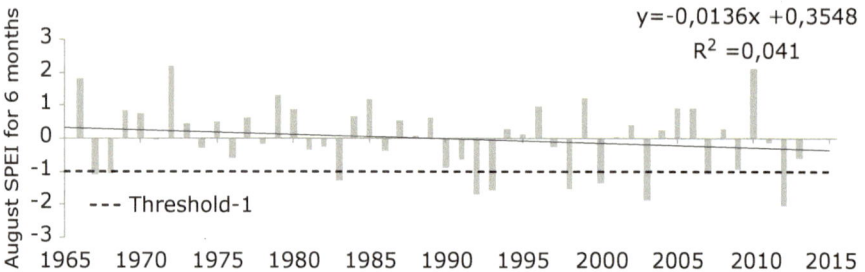

Fig. 2.11 Values of the 6-month SPEI for August with a linear trend of the time series and an indication of the value of the index −1 (reaching/crossing this value indicates severe drought) Vido et al. (2019)

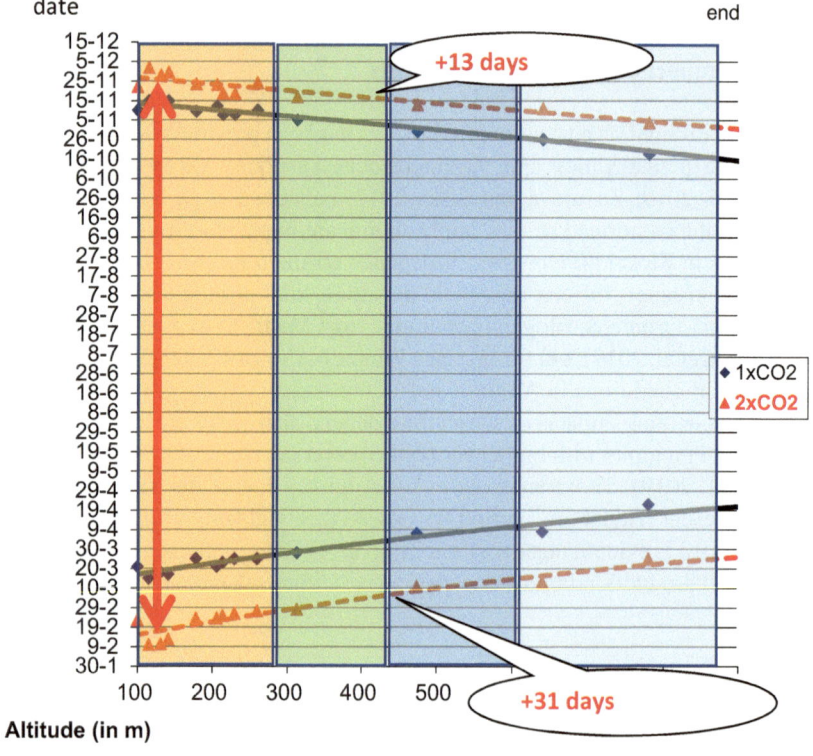

Fig. 2.12 Dependence of onset and end of the vegetative period on altitude for 1xCO₂ and 2xCO₂ climate

duration is changing significantly in evaluated time horizons (Fig. 2.12). The changes in the onset, end, and duration of the vegetation period and changes in climate conditions will have an impact on the zonation of agricultural production and the planting of crops. The sum of active temperatures in the vegetative period (TS5) will probably increase by 22% on lowlands in future climate $2xCO_2$. These sums will increase up to 45% with increasing altitude in mountainous regions.

According to the RCP 4.5 scenario, the expected changes in precipitation during months of the year are not the same. Differences in rainfall are also depended by the altitude. Beside it, the time factor plays an important role. During a longer vegetative period can be more precipitation accumulated. In fact, the increase in precipitation for climate change scenario $2xCO_2$ is by 65–80 mm (15–20%) in lowlands and by 65–128 mm (12–20%) in mountainous regions is supposed.

The increase of air temperature, as well as lengthening of the vegetative period, will cause an increase of potential evapotranspiration (E_0) in the $2xCO_2$ climate. In the corn maize growing areas, the increase of E_0 will be by about 150 mm (23%) in comparison to the reference period. In the mountainous E_0 will increase by 127 mm (30%). E_0 totals exceeding 800 mm could be expected in the warmest areas of Podunajská nížina Valley and the lowest parts of the Východoslovenská nížina Valley. The water demand for field and horticultural crops will increase significantly and a new water management strategy would be highly needed (Šiška and Takáč 2009).

2.2.4.3 Drought Impact on Climatic Water Balance of Agricultural Land

There are many indicators for the evaluation of water availability for primary agricultural production in the landscape. The climate water balance (CWB) indicator has been introduced for purposes of agro-climatic zonation in Slovakia. This indicator is a difference between potential evapotranspiration and precipitation in three summer months (Kurpelová et al. 1975). In fact, summer days occur frequently during spring and autumn months too, so the CWB was evaluated for the whole vegetative period in this study. Climatic water balance (CWB) is significantly changing as the sums of potential evapotranspiration and precipitations are changing in the altitude profile of the Slovak Republic. Šiška and Takáč (2009) found differences in the increase of the indicator by 30% (70 mm) in corn maize growing areas in the $2xCO_2$ climate. The zero values of CWB will move from 550 to 650 m (Fig. 2.13). In the reference period 1961–1990 ($1xCO_2$) there was short of water 21,300 km^2 of agricultural land, in $2xCO_2$ climate is expected increase of vulnerable land by 42% with total area 30,300 km^2. The average deficit E_0-R > 250 mm will occur the most important agricultural areas of Slovakia (Fig. 2.14). Such water shortages were practically absent in the 1961–1990 reference period. Another consequence of climate change is likely to be a diverse response in Slovakia. While in the reference period of 1961–1990 there were defined six areas according to water needs of plant production, at the end of the twenty-first century another two very dry areas should be added.

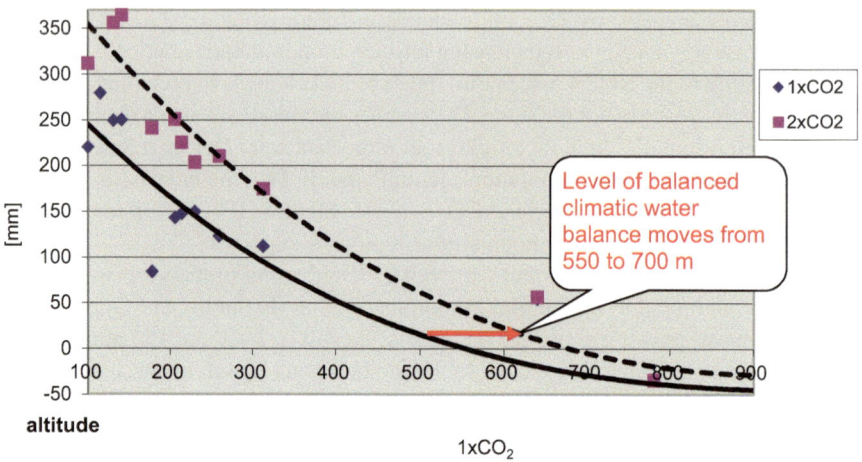

Fig. 2.13 Climatic water balance in the altitude profile of Slovakia for $1xCO_2$ and $2xCO_2$

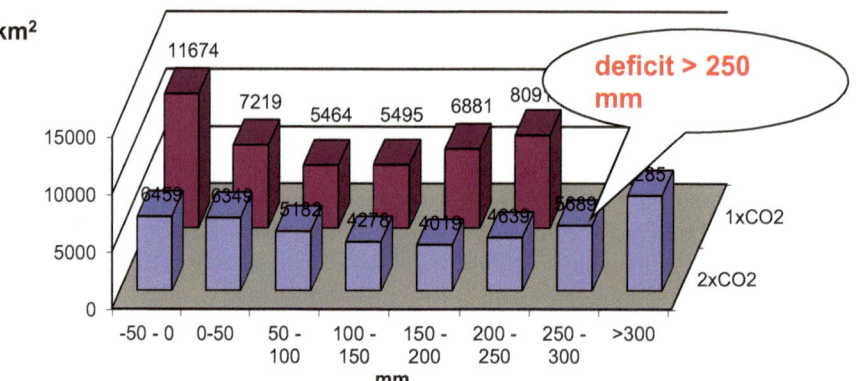

Fig. 2.14 Spatial distribution of CWB in the for $1xCO_2$ and $2xCO_2$

2.2.4.4   Onset, End and Intensity of Drought in Agricultural Productive Zones of Slovakia

The onset of drought and its intensity varies considerably in individual production zones (Šiška and Takáč 2009). The reference climatic period 1961–1990 was characterized by an excess of water in agrocenosis at the beginning of the year in the period of vegetation rest. However, as the vegetation period begins, the situation changed sharply in the warmest corn production zone, when surplus precipitations passed into water scarcity around 10th of March and the cumulative climate water balance reached maximum water scarcity expressed by CWB = −120 to −200 mm in August (Fig. 2.15a). The balanced cumulative climatic water balance was

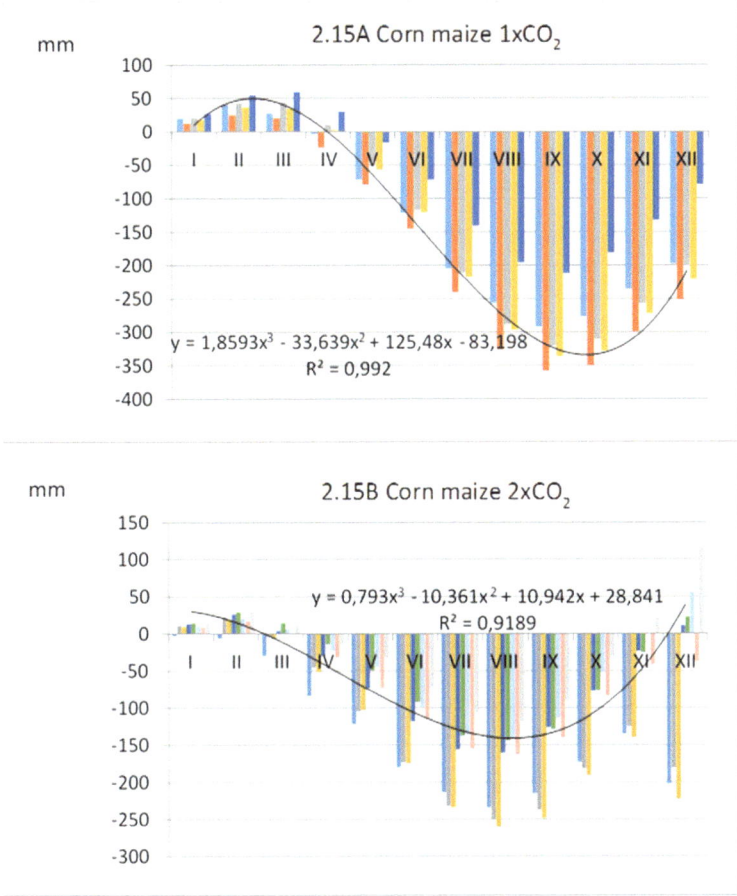

Figs. 2.15 and 2.16   Cumulative climatic water balance (CWB) for corn maize zone (Fig. 2.15) a beet zone (Fig. 2.16) in for $1xCO_2$ (A) $2xCO_2$ (B) in Slovakia

observed again at the end of the year in the lowlands of Slovakia. This trend showed up also in beet production zone (Fig. 2.16a) when the drought was observed in the first days of May with the peak of water scarcity in August. The cumulative climate water balance was CWB = -100 mm and a balanced cumulative climate balance was observed at the end of October. Negative CWB values did not appear in potato and mountain production zone during the reference period (Figs. 2.17a and 2.18a).

The deficit of water occurred from June in the potato production zone. Relatively stable was the only mountain production zone with altitude above 700 m. According to the GCM models, higher precipitation and negative CWB are expected in corn and beet production zones about 12 days later compared to the 1961–1990 reference time period (Figs. 2.15b and 2.16b).

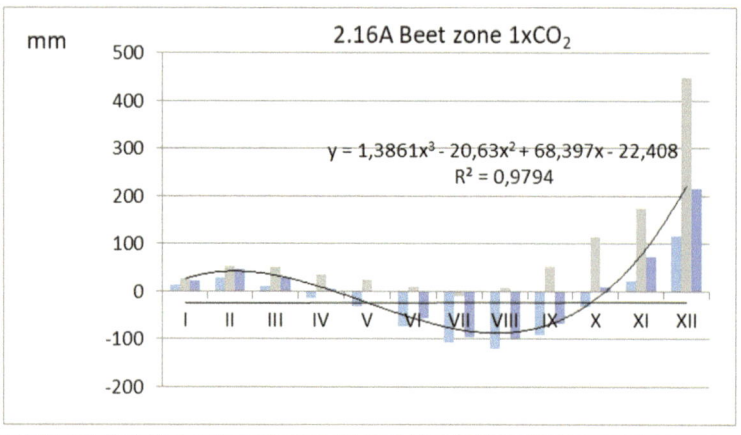

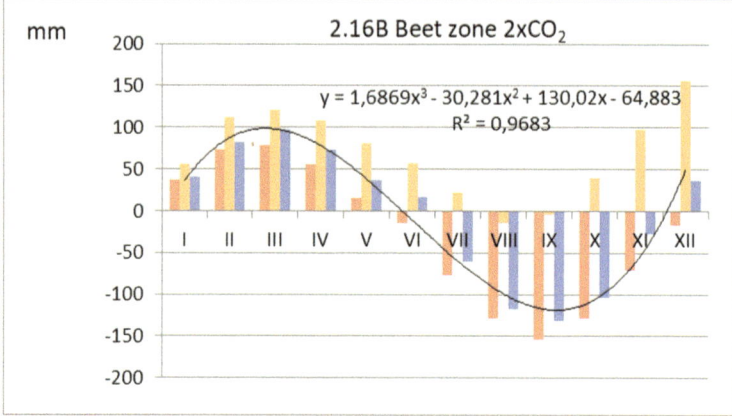

Figs. 2.15 and 2.16 (continued)

2.3   Overview of Drought Assessment Methods Often Used in Agricultural Landscape

2.3.1   Station Based Drought Assessment

Current praxis of drought observation in Slovakia prefer almost in all cases station-based observation methods and indices. It is because of opinion that station-based methods are precise in compare to satellite-based approaches. But the reason is also faster calculation and operations with point data. Frequently used methods are:

(a) Standardised Precipitation Index (SPI)
(b) Standardised Precipitation Evapotranspiration Index (SPEI)
(c) Palmer Drought Severity Index (PDSI)

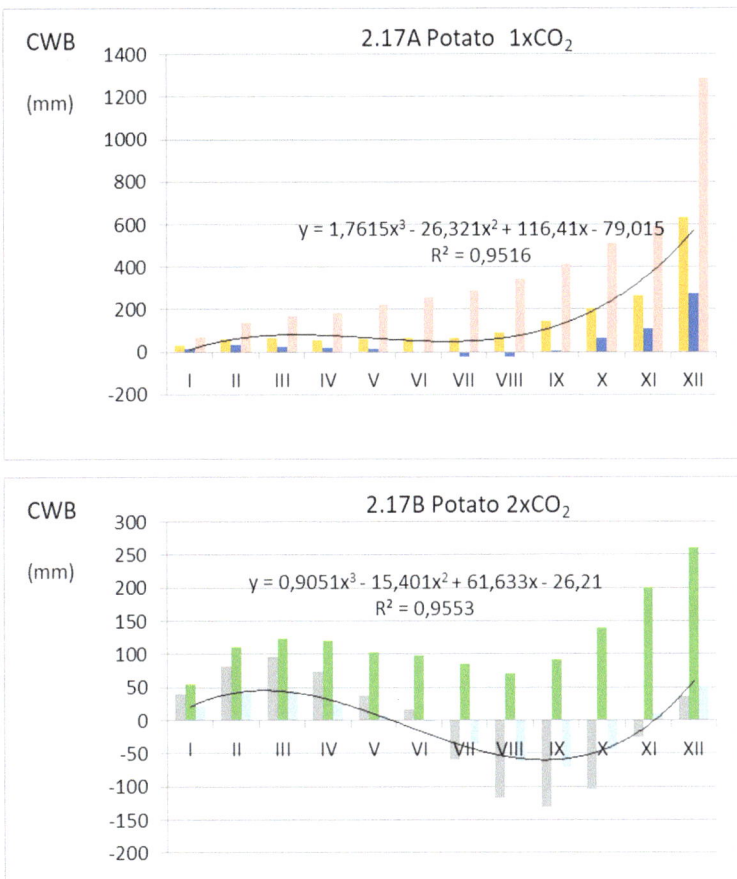

Figs. 2.17 and 2.18 Cumulative climatic water balance (CWB) for potato zone (Fig. 2.17) a mountainous zone (Fig. 2.18) for $1xCO_2$ (A) $2xCO_2$ (B) in Slovakia

Standardized Precipitation Index – is one of the most commonly used drought index worldwide. First introduced in 1993 (McKee et al. 1993). The SPI is based on precipitation data and has the flexibility to detect both short- and long-term precipitation deficits. According to the methodology of McKee et al. (1993), the unique feature of the SPI is that the index can be used to monitor conditions on a variety of time scales. Due to this, it is possible to evaluate diverse impacts of drought from agricultural to hydro-logical sectors. SPI has been in Slovakia used by; Portela et al. (2015), Labudová et al. (2015, 2017), Vido et al. (2015), Šustek and Vido (2013), and Valach et al. (2016).

Standardized Precipitation Evapotranspiration Index – The SPEI is a relatively new drought index introduced by Vicente-Serrano et al. in 2010. The principle of its calculation is based on Standardized Precipitation Index (SPI), which evaluates the

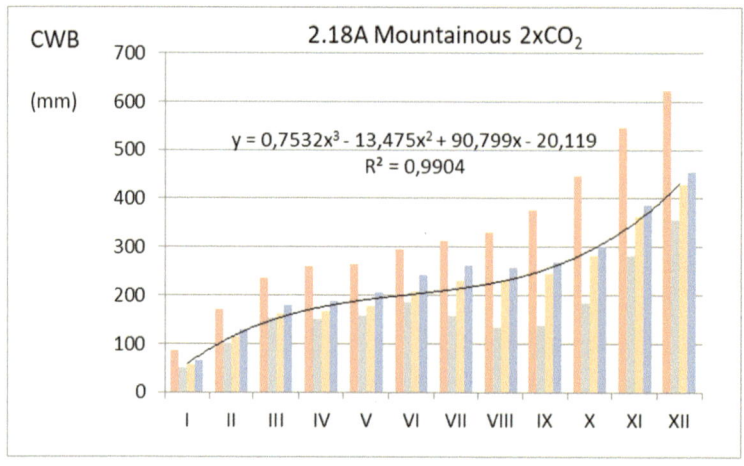

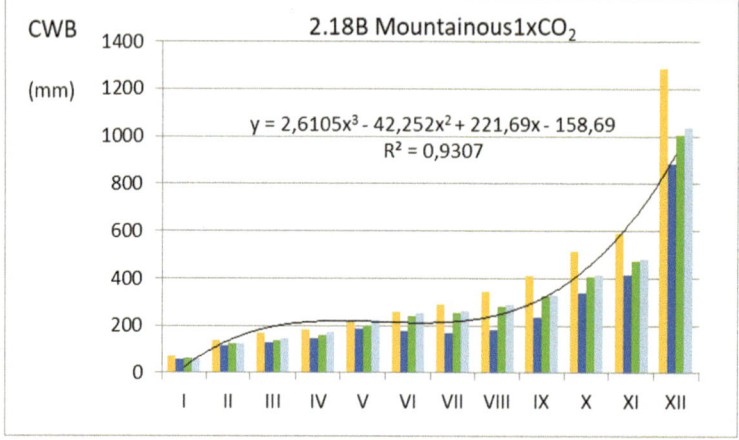

Figs. 2.17 and 2.18 (continued)

deviations of precipitation from the long-term normal at different time scales (usu-ally from 1 to 24 months). One of the limitations of the SPI is that it does not include the passive components of the hydrological regime (i.e., evapotranspiration). Vicente-Serrano et al. (2010) used both precipitation and potential evapotranspira-tion (PET) to generate the SPEI values that include the deviation of the whole cli-matic balance (P-PET) from the normal (i.e., positive values represent positive balance, and vice versa). Following the methodology of Vicente-Serrano et al. (2010), a drought episode starts (similar to the SPI methodology) when a negative value of the index appears and lasts until the first positive value. Because of this, SPEI become very popular also in Slovak research community (Vido et al. 2014, 2019; Labudová et al. 2017; Šustek et al. 2017).

Palmer Drought Severity Index – Developed in the 1960s as one of the first attempts to identify droughts using more than just precipitation data. Palmer was tasked with developing a method to incorporate temperature and precipitation data with water balance information to identify droughts in crop-producing regions of the United States. For many years, PDSI was the only operational drought index, and it is still very popular around the world. Developed mainly as a way to identify droughts affecting agriculture, it has also been used for identifying and monitoring droughts associated with other types of impacts. In Slovak conditions Zuzulová and Šiška (2017), Trnka et al. (2016), Büntgen et al. (2010) and Tall (2008) dealt with PDSI in drought analyses.

2.3.2   Satellite Based Assessment

The first researchers of plant spectral properties were Willstätter and Stoll (1913), who in their study of chlorophyll investigated the entering of light into plant cells (Porra 2002; Mróz and Sobieraj 2004).

The satellite remote sensing also serves to obtain physical and physiological parameters of vegetation. Some Earth Observing System (EOS) tools are targeted to measure the reflected solar radiation by vegetation at certain wavelength intervals. In remote sensing, broadband red (0.6–0.7 μm) radiation and near-infrared (0.75–1.35 μm) radiation are most important. Typically, the measured spectral reflectance data are inserted into vegetation indices (Myneni et al. 1995).

Vegetation indices are indicators describing vegetation (relative density and health status) on satellite imagery. The data from satellite images are transformed into vegetation indices by mathematical formulas – algorithms (U.S. Geological Survey 2015).

Vegetation indices derived from satellite imagery data are one of the basic information sources for monitoring vegetation and its changes. They are usually obtained from spectral reflectance in red and near-infrared channels, based on which vegetation cover is evaluated (Kasawani et al. 2010). Also, the contrast between vegetation and soil is the highest in these channels (Leblon 1993).

Tucker et al. (1985) report that the widespread use of vegetation indices is due to the possibility of their application for classification of vegetation cover in large areas, e.g. on a continental level and on a multitemporal scale.

Vegetation indices can be divided into two basic categories:

(a) Ratio-based indices, that give a simple or normalized ratio of surface reflectance in the red and near-infrared portions of the spectrum. They represent a suitable tool for determining the health of vegetation, yields of cultivated crops, changes in time, course of phenophases, water stress of vegetation, etc. There may be a significant correlation between ratio-based indices and some other vegetation indicators (Dobrovolný 1998). One of the indicators is e.g. Leaf Area Index (LAI), which is defined as the total one-sided area of photosynthetic leaf tissue per unit area (ground surface) (Gobron 2008). The category indices include e.g. RVI, NDVI, and others.

(b) Orthogonal-based indices that represent linear combinations of original bands on a multispectral image. If the original bands create a multidimensional space, by the appropriate rotation of its coordinates it is possible to emphasize certain information on the original image, e.g. vegetation component. From the perspective of the spectral reflectance, one pixel does not only inform about a homogeneous surface but also provides spectrally mixed information. For instance, agricultural land includes the spectral reflectance of the vegetation cover as well as the soil substrate. By using orthogonal-based indices, the reflectivity of vegetation can be separated to some extent from soil reflectivity (Dobrovolný 1998). These are e.g. PVI and others.

Furthermore, the vegetation indices can be divided into:

(a) general, resp. basic (RVI, NDVI, and others)
(b) corrected in terms of atmospheric influence (EVI, GEMI, and others)
(c) corrected in terms of soil influence (SAVI, TSAVI, MSAVI, OSAVI, GESAVI, and others).

2.3.2.1 RVI (Ratio Vegetation Index)

RVI is probably the first implemented vegetation index. It was defined by Pearson and Miller (1972). It is the simplest index of vegetation indices (Jackson and Huete 1991). RVI is expressed by the following formula:

$$RVI = \frac{NIR}{R} \qquad (2.1)$$

where

NIR – spectral reflectance in the near-infrared channel,
R – spectral reflectance in the red channel.

2.3.2.2 NDVI (Normalized Difference Vegetation Index)

The normalized difference vegetation index was introduced in 1973 (Rouse et al. 1973). It has become the most widely used vegetation index, which confirms many scientific papers from various areas using this method to evaluate the vegetation cover. For instance, in Forestry Bucha et al. (2011), Brandýsová and Bucha (2012) used NDVI to monitor the phenological phases of beech stands; Moghaddam et al. (2015) identified a forest fire in Golestan; Maselli (2004) evaluated the conditions of coniferous and broadleaved forests in Tuscany; Meneses-Tovar (2009) determined the degradation of forest vegetation in Mexico. In the field of agriculture, this index is used to estimate and predict yields, e.g. rice (Huang et al. 2013) wheat (Zhang et al. 2012; Sultana et al. 2014), corn (Rojas 2013) and other crops. NDVI can be used for monitoring drought (Peters et al. 2002; Karnieli et al. 2010; Berhan et al. 2011). In the field of nature conservation, Biswal et al. (2013) analyzed the temporal and spatial changes in the density of vegetation cover in the Similipal Biosphere Reserve in Odisha, India.

Values are expressed in a range from 1 to −1. Rocky surfaces, sand, snow cover, and barren areas usually have less value than 0.1. Sparse vegetation such as e.g. shrubs or meadows and pastures are expressed by values of about 0.2–0.5. High values from about 0.6 to 0.9 belong to areas with dense vegetation covers such as e.g. forest or cultivated crops in the most intense growth phase (U.S. Geological Survey 2015). The formula is presented in Chap. 3.

2.3.2.3   PVI (Perpendicular Vegetation Index)

PVI was first used by Richardson and Wiegand in 1977. It represents an orthogonal-based vegetation index, an index based on the perpendicular distance between the soil line and the vegetation pixels. In the case of vegetation, the reflectance is high in the near-infrared band, which means that the vegetation pixel will always be above the soil line (Panda et al. 2010). PVI is expressed by the following formula:

$$PVI = \sqrt{\left(NIR_S - NIR_V\right)^2 + \left(R_S - R_V\right)^2}$$ (2.2)

where

NIR_S – spectral reflectance of soil in the near-infrared channel,
NIR_V – spectral reflectance of vegetation in the near-infrared channel,
R_S – spectral reflectance of soil in the red channel,
R_V – spectral reflectance of vegetation in the red channel.

2.3.2.4   SAVI (Soil-Adjusted Vegetation Index)

The SAVI (Soil-Adjusted Vegetation Index) was developed from the normalized difference vegetation index equation to limit the effects of soil on the vegetation cover spectrum by introducing a soil correction factor (L). This factor should change inversely with the amount of vegetation in a certain area. This means, the greater the amount of vegetation, the lower the factor L value (Qi et al. 1994). SAVI is calculated based on the following formula:

$$SAVI = \frac{NIR - R}{NIR + R + L} \times \left(1 + L\right)$$ (2.3)

where

L – soil correction factor, usually 0.5,
NIR – spectral reflectance in the near-infrared channel,
R – spectral reflectance in the red channel.

Huete (1988), who proposed the use of the soil correction factor L, explored its optimal value. The optimal value of L varies according to the density of the vegetation, and thus there may be two or even three optimal values. For a very sparse vegetation cover, it is a value of 1, for a medium-density vegetation cover, it is a value of 0.5 and a value of 0.25 for a higher density vegetation cover. By introducing an L factor of 0.25–1, compared to NDVI or PVI the soil influence was significantly reduced.

2.3.2.5   TSAVI (Transformed Soil-Adjusted Vegetation Index)

TSAVI, like SAVI, was introduced to eliminate errors in measuring vegetation cover with very low density. However, this index measures the angle between the soil line, which is a linear regression of the near-infrared and red bands on bare soil pixels and a line that connects the vegetation points and points associated with the soil line. It was proposed by Baret et al. already in 1989, but in 1991 he improved it to the version provided hereby (Baret et al. 1989; Baret and Guyot 1991):

$$TSAVI = \frac{A \times (NIR - A \times R - B)}{\left[A \times NIR + R - A \times B + 0.08 \times \left(1 + A^2\right) \right]} \qquad (2.4)$$

where

A, B – soil line parameters,
0.08 – a value that minimizes the impact of background soil brightness,
NIR – spectral reflectance in the near-infrared channel,
R – spectral reflectance in the red channel.

If the output TSAVI value is 0, it means that it is a land without vegetation cover. If the value is close to 0.70, then the area of interest is covered with dense vegetation (Baret and Guyot 1991).

2.3.2.6   MSAVI (Modified Soil-Adjusted Vegetation Index)

Qi et al. (1994) modified the SAVI equation by replacing the L-correction coefficient by L function. Compared to SAVI, which has a very mild sensitivity to soil influence, MSAVI almost entirely removes soil impacts. Through the procedure specified in their work, they have reached a final formula:

$$MSAVI = \frac{2NIR + 1 - \sqrt{(2NIR + 1)^2 - 8(NIR - R)}}{2} \qquad (2.5)$$

where

NIR – spectral reflectance in the near-infrared channel,
R – spectral reflectance in the red channel.

2.3.2.7 OSAVI (Optimized Soil-Adjusted Vegetation Index)

OSAVI was introduced in 1996 by Rondeaux et al. (1996). The disadvantage of this index is that it targets the vegetation cover generally. This means that it is not possible to differentiate healthy vegetation from the damaged one because it cannot be used for all spectral data, just a few bands (Steven 1998). OSAVI is determined by the following formula:

$$OSAVI = \frac{NIR - R}{NIR + R + 0.16}$$ (2.6)

where

0.16 – value that minimizes the impact of soil,
NIR – spectral reflectance in the near-infrared channel,
R – spectral reflectance in the red channel.

2.3.2.8 GESAVI (Generalized Soil-Adjusted Vegetation Index)

GESAVI was developed in 1998 (Gilabert et al. 1998, cit. Gilabert et al. 2002). This index is based on the angular distance between the soil line and the vegetation isolines. Its sensitivity to soil background is lower, which makes it a very effective index. This results from the fact that vegetation isolines in the NIR-R are not parallel to the soil line (as in the case of PVI) and are neither convergent at the origin (as in NDVI) but converge somewhere between the origin and the infinity in the negative values in both spectrum channels. It is defined by the following formula:

$$GESAVI = \frac{NIR - AR - B}{R + Z_S}$$ (2.7)

where

A, B – soil line parameters,
Z_S – soil correction coefficient related to reflectance in the red channel at the intersection of the soil line and the vegetation isolines, value 0.35,
NIR – spectral reflectance in the near-infrared channel,
R – spectral reflectance in the red channel.

This vegetation index was tested using a set of reflectivity data in laboratory conditions and then applied to Landsat TM satellite imagery to determine vegetation in semiarid regions. The resulting GESAVI value varies between 0 and 1, where 0 represents soil without vegetation, and values close to 1 represent dense vegetation cover (Gilabert et al. 2002).

2.3.2.9   EVI (Enhanced Vegetation Index)

This index is calculated similarly to NDVI, but it allows to adjust the distortion caused by different particles in the air as well as the soil cover under vegetation stand (Weier and Herring 2000; Huete et al. 2002; Jiang et al. 2008). Defined by the following formula:

$$EVI = 2.5\frac{NIR - R}{NIR + C_1 \times R - C_2 \times Blue + L} \tag{2.8}$$

where

NIR – atmospherically corrected spectral reflectance in the near-infrared channel,
R – atmospherically corrected spectral reflectance in the red channel,
$Blue$ – atmospherically corrected spectral reflectance in the blue channel,
C_1, C_2 – aerosol resistance coefficients over time using the blue spectrum to correct the effect of aerosols in the red channel; $C_1 = 6$ a $C_2 = 7.5$,
L – the correction factor for soil whose value is 1, i.e. higher than in SAVI, due to interactions and feedbacks between L and aerosol resistance term.

2.3.3   Combined Station/Satellite Based Drought Assessment

One of the most frequent station/satellite combined method of drought monitoring is VegDRI (Vegetation Drought Response Index). VegDRI was introduced in 2008 as a new integrated (combination of station based observations and satellite based monitoring) approach for monitoring of drought stress in vegetation (Brown et al. 2008). VegDRI is based on composite of ground climatic (weather) observations, satellite observations of vegetation response (NDVI) and biophysical data, as ecological regions, land cover or land use, percent of irrigated area and soil conditions represented as average water holding capacity. It is highly desirable to use all available data because as known drought is shape of climate, but impacts are highly influenced by local environmental conditions. Therefore, higher number of potential factors increases quality of VegDRI outputs. The VegDRI index could identified and estimate drought impact on vegetation due to its composite structure based on ground climate observation and satellite vegetation response observation (NDVI). Thus, this tool could be very useful in operative estimation and monitoring drought within the region. However this method depend on various data availability (short period of record due to remotely sensed data). Also density of meteorological stations influence the quality of the output from the VegDRI (Fig. 2.19). Despite of this challenges, it is no doubt that this combined approach is the future of the drought monitoring.

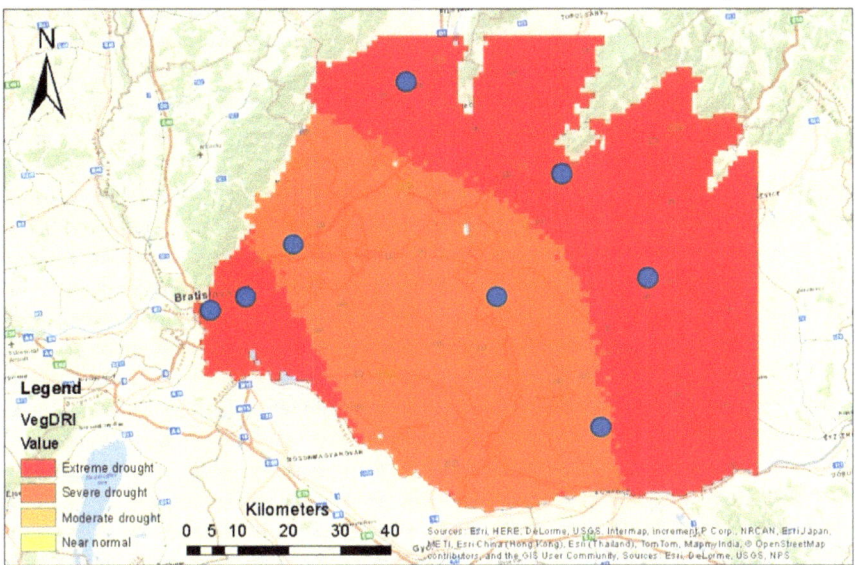

Fig. 2.19 Illustrative map of the VegDRI experimental operation in Podunajská nížina Valley for the first period of August 2003. Map based on observation of eight meteorological stations

Chapter 3
Material and Methods

Abstract The chapter consists of a description of the method of drought assessment by the PDSI, processing of remote sensing imagery and derivation of the NDVI and assessment of drought impact on yields of spring barley (*Hordeum vulgare* L.), winter wheat (*Triticum aestivum* L.) and maize (*Zea mays* L.). Hectare yields of the cereals were detrended for each district using the second degree polynomial function. For identification of relationships between PDSI and NDVI and drought impact on vegetation vigor was used correlation analysis.

Keywords PDSI · NDVI · Yields · Correlations

3.1 Climate Data Analyses and Drought Indices

Drought assessment was realized by calculation of monthly values of Palmer Drought Severity Index (PDSI), which was chosen among other drought indices. PDSI was developed in the second half of the 1960s by Palmer (1965). It is standardized for different regions with different climatic conditions (Dunkel 2009). In the past, e.g. climatic index of drought (ETP – P in mm) was used for the assessment of drought in Slovakia (Šiška and Takáč 2009; Kurpelová et al. 1975).

Palmer in his index of drought used Thornthwaite's method of potential evapotranspiration calculation, which has the advantage that it is based solely on the average monthly air temperature (Palutikof et al. 1994). In comparison with the newer methods, the disadvantage is the lower accuracy because newer methods use multiple elements (Litschmann et al. 2002). Many authors replace this equation with the Penman-Monteith equation which is more profound for input data (e.g. Tall and Gomboš 2011; Litschmann et al. 2002; Litschmann and Rožnovský 2001).

© The Author(s), under exclusive license to Springer Nature Switzerland AG 2020
V. Zuzulová et al., *Agricultural Drought in Slovakia: An Impact Assessment*,
SpringerBriefs in Environmental Science,
https://doi.org/10.1007/978-3-030-42061-1_3

Thornthwaite's potential evapotranspiration formula (1948):

$$ETP = 1.6\left(\frac{10T}{I}\right)^{a} \ [\text{mm}] \tag{3.1}$$

where

ETP – potential evapotranspiration in mm per month,
T – average monthly air temperature in °C.

$$I = \sum_{1}^{12}\left(\frac{T_i}{5}\right)^{1.51} \tag{3.2}$$

where

I – heat index,
T_i – long-term average air temperature in the given month of the year in °C.

$$a = 0.675 \times 10^{-6} \times I^3 + 77.11 \times 10^{-6} \times I^2 + 17.921 \times 10^{-3} \times I + 0.49239 \tag{3.3}$$

The following four coefficients are needed to calculate the water balance on a monthly basis (Alley 1984). In these calculations, the soil profile is divided into two layers. The top layer is assumed to contain 25.4 mm (1 in.) of available water capacity. The lower layer has a value of available water capacity depending on soil characteristics and depth of balancing. In their work, Litschmann and Rožnovský (2001) consider balancing depths up to 60 cm, Tall and Gomboš (2011) with up to 100 cm balancing depth.

$$\alpha_i = \frac{\overline{ET_i}}{\overline{ETP_i}} \tag{3.4}$$

$$\beta_i = \frac{\overline{R_i}}{\overline{PR_i}} \tag{3.5}$$

$$\gamma_i = \frac{\overline{RO_i}}{\overline{PRO_i}} \tag{3.6}$$

$$\delta_i = \frac{\overline{L_i}}{\overline{PL_i}} \tag{3.7}$$

where

ET_i – evapotranspiration for a month,
ETP_i – potential evapotranspiration for a month,
R_i – recharge; net gain in soil moisture during the month

PR_i – potential recharge; the amount of moisture needed to saturate the soil profile to field capacity for an individual month,

RO_i – runoff for a month,

PRO_i – potential runoff, i.e. potential precipitation minus potential recharge; Palmer expects potential precipitation to be equivalent to available water capacity,

L_i – loss of soil moisture during a month,

PL_i – potential loss; the amount of moisture that can be lost from the soil to evapotranspiration in the event of zero precipitation in a month.

They express the ratio of average actual and potential values of variables in an individual month for a selected location. The units of measurement are inches.

The long-term precipitation average for a location is then calculated:

$$\hat{P} = \alpha_i ETP + \beta_i PR + \gamma_i PRO - \delta_i PL \tag{3.8}$$

Then the moisture departure (d) for each month is calculated as a difference between actual and calculated monthly precipitation. It represents an excess or lack of precipitation (P) compared to calculated (climatically appropriate for existing conditions) precipitation ($\hat{P}$).

$$d = P - \hat{P} = P - \alpha_i ETP + \beta_i PR + \gamma_i PRO - \delta_i PL \tag{3.9}$$

The next calculation is the calculation of K, the climate characteristic, which is calculated in the following way:

$$K_i = \left(\frac{17.67}{\sum\limits_{i=1}^{12} \overline{d}_i K'_i} \right) K'_i \tag{3.10}$$

where

$\overline{d}_i$ – average monthly moisture departure

$$K'_i = 1.5 \log_{10} \left(\frac{\dfrac{\overline{ETP}_i + \overline{R}_i + \overline{RO}_i}{\overline{P}_i + \overline{L}_i} + 2.8}{\overline{d}_i} \right) + 0.5 \tag{3.11}$$

Multiplication of the moisture departure (d) and climatic characteristic (K) results in a moisture anomaly index or Z index:

$$Z = dK \tag{3.12}$$

However in order to satisfy the condition that a rare occurrence of a wet month does not mean the end of a dry season (and vice versa), as well as a period with normal precipitation totals following a dry (wet) season, this equation is calculated for three different indices X_1, X_2 a X_3.

X_1 – severity Index for a wet spell that is becoming established,
X_2 – severity Index for a drought that is becoming established,
X_3 – severity Index for any wet spell or any drought that has become established.

The index X_1 is always positive, while X_2 is always negative or equal to zero. When these limits are exceeded, the respective index always resets and its calculation starts again. The dry season starts when the value of the index $X_2 \leq -1$, wet if $X_1 \geq 1$. All this, provided that the previous dry or season period has ended. In these cases, the index value $X_3 = X_1$ for the wet season and $X_3 = X_2$ for the incoming dry season. Then one of these three indices are selected for the PDSI in the individual month (Litschmann et al. 2002).

The identification of the Palmer Drought Severity Index in the present work was calculated through the program provided by the Slovak Hydrometeorological Institute in Bratislava. The program was developed at UNL (University of Nebraska – Lincoln) July 15, 2003. It is written in Fortran. The author of the program is Tom Heddinghaus. The input data for the calculation consist of average monthly precipitation totals, average monthly air temperature, average air temperature of the whole period, latitude and available water capacity. Climatic data were provided by Slovak Hydrometeorological Institute (SHMI) and data of available water capacity by Soil Science and Conservation Research Institute (SSCRI).

Twelve meteorological stations were selected for assessment of drought in Slovakia: Bratislava, Piešťany, Hurbanovo, Čadca, Sliač, Boľkovce, Rimavská Sobota, Telgárt, Poprad, Košice, Milhostov and Kamenica nad Cirochou (Fig. 3.1).

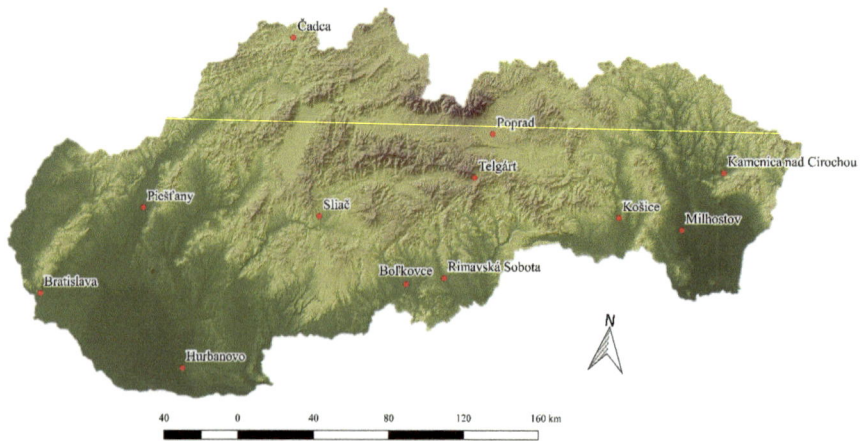

Fig. 3.1 Meteorological stations selected for drought assessment: Bratislava, Piešťany, Hurbanovo, Čadca, Sliač, Boľkovce, Rimavská Sobota, Telgárt, Poprad, Košice, Milhostov, Kamenica nad Cirochou

Table 3.1 Classification based on the Palmer Drought Severity Index

PDSI	Class
≥ 4.00	Extremely wet
3.00 to 3.99	Very wet
2.00 to 2.99	Moderately wet
1.00 to 1.99	Slightly wet
0.50 to 0.99	Incipient wet spell
0.49 to −0.49	Near normal
−0.50 to −0.99	Incipient drought
−1.00 to −1.99	Mild drought
−2.00 to −2.99	Moderate drought
−3.00 to −3.99	Severe drought
≤ −4.00	Extreme drought

Source: Palmer (1965)

These stations were selected based on limited climatic parameters. The assessment was made for 55-year time series from 1960 to 2014.

As dry seasons of the period were identified those with values ≤ −1 and at least 1 month during the season was classified as moderately dry (−2.00 to −2.99) (Žalud et al. 2006) according to the Palmer classification (Table 3.1).

3.2 Field Crops Data Analyses

As model crops were selected spring barley (*Hordeum vulgare* L.), winter wheat (*Triticum aestivum* L.) and maize (*Zea mays* L.) because of their spread in Slovakia and the varied demand of these crops for moisture in time. Spring barley uses spring precipitation, winter wheat uses winter supplies and maize uses summer precipitation.

Hectare yields of selected cereals for individual districts in time series of 2000–2014 were provided by the Statistical Office of the Slovak Republic (2019). The data were collected for 71 districts, as the districts of Bratislava I–V were evaluated together as well as the districts of Košice I–IV and Košice – okolie were evaluated as a whole.

Due to insufficient quality of data, districts in which data for the entire time series were not recorded were excluded. In the case of spring barley, the following four districts were not evaluated: Čadca, Kysucké Nové Mesto, Žiar nad Hronom and Medzilaborce. In the case of winter wheat Čadca district was excluded. Concerning to maize, 30 districts were excluded from the evaluation: Považská Bystrica, Bytča, Čadca, Dolný Kubín, Kysucké Nové Mesto, Liptovský Mikuláš, Martin, Námestovo, Ružomberok, Turčianske Teplice, Tvrdošín, Žilina, Banská Bystrica, Banská Štiavnica, Brezno, Detva, Žarnovica, Žiar nad Hronom, Humenné, Kežmarok, Levoča, Medzilaborce, Poprad, Sabinov, Snina, Stará Ľubovňa, Stropkov, Svidník, Gelnica and Spišská Nová Ves.

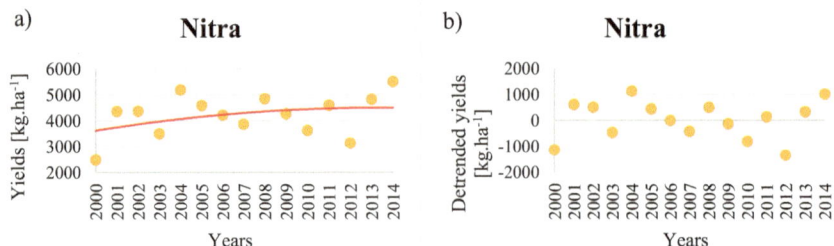

Fig. 3.2 Yields of spring barley [kg.ha⁻¹] (**a**) and detrended yields of spring barley by second degree polynomial function (**b**) in Nitra district during the period 2000–2014

The original hectare yields were detrended for each district and each cereal (as it is shown by example of the spring barley yield in the district of Nitra – Fig. 3.2), because of a significant trend of the crop yield fluctuation, which could be caused by the change of in cultivation practices (e.g. intensive plant protection, fertilization). In total, 178 trends were graphically evaluated, which were subsequently detrended using the second degree polynomial function.

3.3   Satellite Data

The vegetation vigor assessment was realized by normalized difference vegetation index (NDVI), the calculation of which is expressed by the following formula:

$$NDVI = \frac{NIR - R}{NIR + R} \tag{3.13}$$

where

NIR – spectral reflectance in the near-infrared channel,
R – spectral reflectance in the red channel.

The NDVI value is affected by the absorption of the red band of the spectrum by the leaf pigments and the reflectance of the leaves in the near-infrared band of the spectrum. The development of the assimilation apparatus causes a decrease in the reflectance in the red band of the spectrum while increasing the reflectivity in the infrared band of the spectrum. This means that in the case of green vegetation, NDVI is increased (Koch et al. 1990).

The images from the remote sensing were made by MODIS (Moderate Resolution Imaging Spectroradiometer), one of the five instruments of Terra satellite, which belongs to Earth Observation System (NASA 2016a). MODIS is an instrument that provides multi-spectral observations of the Earth's surface, atmosphere features, and oceans at a spatial resolution from 250 to 1000 m. It provides about 40 data

Fig. 3.3 Image with
NDVI values multiply by
scale factor 0.0001.
(Source: NASA,
EOSDIS 2016)

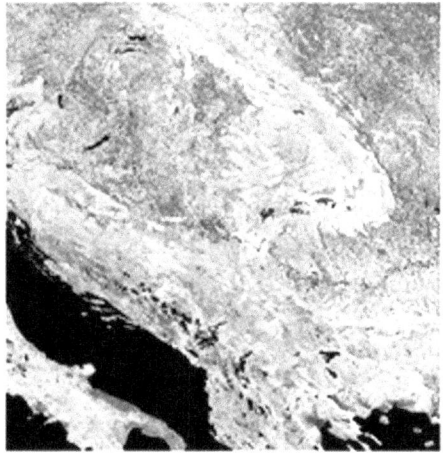

Fig. 3.4 Image with
NDVI values relevant to
vegetation cover. (Source:
Own processing (based on
initial input data from
NASA, EOSDIS
2016), 2016)

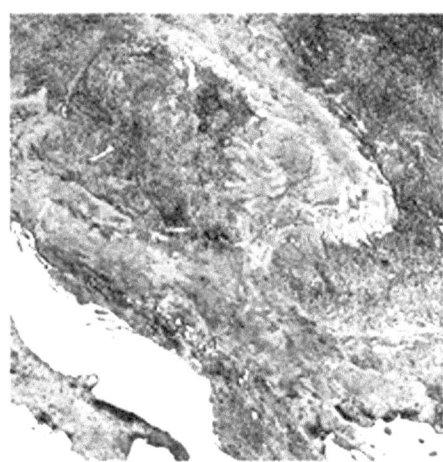

products, including vegetation indices (NASA 2016b). For the purpose of this dissertation, NDVI was derived from MOD13Q1, available from reverb.echo.nasa.gov. The images spatial resolution was 250 m and the temporal resolution was 16 days. The total number of processed images was 357. The NDVI was evaluated in the time series from 2000 to 2014.

Working with images was done according to the following steps (Figs. 3.3, 3.4, 3.5 and 3.6):

1. Conversion of the image format in the MODIS Reprojection Tool (MRT) (LP DAAC, U.S. Geological Survey, EROS 2011) from the original hdf (in which the images are distributed) to tif format, which is suitable for processing in ArcGIS.

ArcGIS Desktop is an American software by ESRI, Inc. It consists of a set of products for creating, managing, analyzing, and visualizing geodata that enables the building of the geographic information system. ArcGIS is licensed in three

Fig. 3.5 NDVI of vegetation cover on arable land in Slovakia. (Source: Own processing (based on initial input data from NASA, EOSDIS 2016), 2016)

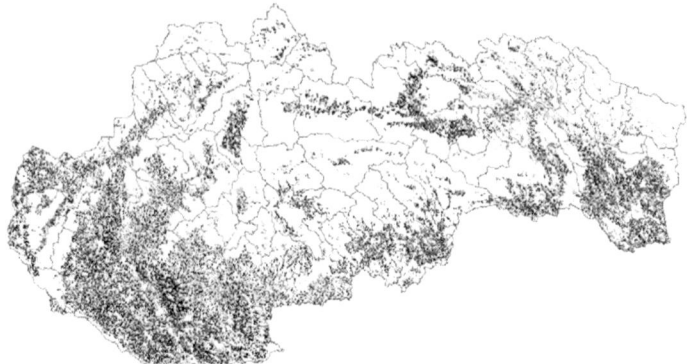

Fig. 3.6 NDVI of vegetation cover on arable land in each district in Slovakia. (Source: Own processing (based on initial input data from NASA, EOSDIS 2016), 2017)

levels - ArcView, ArcEditor, and ArcInfo, which are different in their functionality. ArcView is a set of applications that are ArcMap, ArcCatalog, ArcToolbox, and ModelBuilder. ArcView is a basic tool for creating maps, for retrieving information from maps using basic map analysis and editing data in a shapefile format or in a geodatabase. ArcEditor is a solution for retrieving, editing and managing geographic data in personal geodatabases and enterprise geodatabases. Includes tools for metadata creation, an extended offer of tools for work with geographic data and mapping tools, and advanced cartographic tools. Unlike ArcView, it allows definition and administration of geodatabases (topography, subtypes, domains, geometric networks) as well as their editing and compilation. ArcInfo is the highest level in terms of functionality, from the three described license levels. It includes all the capabilities of ArcView and ArcEditor, plus advanced data processing through ArcToolbox. ArcMap is the most widely used application from ArcGIS. It enables creation and editing of spatial data, analysis of these data and also the visualization.

It offers a number of features for map-based tasks (Základní informace – ArcGIS Desktop, 2014).

2. Multiply by a scale factor, which is 0.0001 for NDVI (LP DAAC, NASA, U.S. Geological Survey 2016).
3. Reduction of NDVI values without vegetation cover (< 0.3), e.g. water surfaces, bare rocks, sands, snow cover.
4. Application of layer with arable land in Slovakia.
5. Calculation of monthly NDVI values by averaging two temporally adjacent 16-day NDVI values and averaging monthly NDVI values in each district in a selected time series.

3.4 NDVI and PDSI Methodology

The calculations of correlations between NDVI values and PDSI values during the main growing season limited by the temperatures of ≥ 10 °C (84 relationships were evaluated), and correlations between detrended yields of spring barley, winter wheat and maize in kg.h^{-1} and NDVI values (a total of 1103 relationships were evaluated) were performed in Statgraphics and Microsoft Excel programmes. For the quantification of relations, Pearson's correlation coefficient (r) was calculated. It was developed by the English mathematician and philosopher Karl Pearson in 1896. It is used for the determination of the strength of the statistical dependence of two quantitative variables. The Pearson correlation coefficient is calculated according to the formula:

$$r = \frac{\Sigma(x-\bar{x})(y-\bar{y})}{\sqrt{\Sigma(x-\bar{x})^2 \, \Sigma(y-\bar{y})^2}} \tag{3.14}$$

The numerator (covariance) expresses how the values of two variables change simultaneously. A positive value indicates that variables change in one direction, a negative value indicates that they change in the opposite direction. If the numerator remains zero, it means that the values change independently. The resulting value of Pearson's correlation coefficient ranges from −1 to 1. The value −1 indicates that all observations are on a decreasing line and the value 1 indicates that the observations are on a rising line. The conclusions are not determined solely by the coefficient value, it is necessary to examine the X-Y graph (dependence) as it can be greatly influenced by extreme values. A strong non-linear relationship between two variables can also be determined from the graph (Rimarčík 2014).

The calculation of the coefficient of determination (r^2) was realized, which represents the proportion of the common variability between two variables (Dufour 2011).

$$r^2 = \frac{S_e}{S_T} \tag{3.15}$$

where

S_e – explained variance,
S_T – total variance.

P-values (level of significance) were also calculated, that expresses the probability that the dependence between variables is random.

Chapter 4
Results

Abstract Results aimed at evaluating relationships between NDVI (Normalized Difference Vegetation Index) and PDSI (Palmer Drought Severity Index) as predictors of crop yield are presented in the chapter. As the most important result is the potential of application of NDVI as a relatively reliable crop yield forecast predictor. Because this prediction method is statistical, a full overview of the correlations between the crop yields for specific months is provided. At the end of the chapter, the potential of practical application for end users outlined.

Keywords NDVI · PDSI · Application · End users · Crop yields · Statistics · Correlation analyses

4.1 Evaluation of Relationships Between NDVI and PDSI

The relationship between vegetation vigor estimating by NDVI and a drought estimating by PDSI was determined from the average monthly values of both indices. The data were calculated for the monitored sites and the 12 districts to which they belong during the main growing season (April–October) in the period 2000–2014.

In case of Pearson's correlation coefficient (Table 4.1) we used blue colour to highlight low correlated variables ($0.3 \leq r < 0.5$); green colour to highlight moderately correlated variables ($0.5 \leq r < 0.7$) and red colour was used to highlight highly correlated variables ($0.7 \leq r < 0.9$). Variables which can be considered as very highly correlated ($0.9 \leq r$) was not observed. Results showed that the correlations between vegetation vigor and drought from low to high on the sites were seen mainly from June to September. Share of these correlations from all selected sites represented 92% in June, also 92% in July, 100% in August and 83% in September. In April, May, and October the negative values were observed. It means that increasing PDSI is associated with decreasing NDVI, and vice versa. It was probably caused by phenophases of crops during the growing season.

© The Author(s), under exclusive license to Springer Nature Switzerland AG 2020
V. Zuzulová et al., *Agricultural Drought in Slovakia: An Impact Assessment*, SpringerBriefs in Environmental Science, https://doi.org/10.1007/978-3-030-42061-1_4

Table 4.1 Pearson's correlation coefficient (r) between vegetation vigor (NDVI) and drought (PDSI) on the selected sites from April to October during the period 2000–2014

Sites (stations)	Months						
	IV.	V.	VI.	VII.	VIII.	IX.	X.
Bratislava	−0.25	0.32	0.67	0.62	0.55	0.24	0.24
Piešťany	−0.33	0.17	0.47	0.49	0.63	0.65	0.08
Hurbanovo	−0.53	0.20	0.28	0.52	0.62	0.50	−0.06
Čadca	0.43	0.39	0.52	0.22	0.48	0.74	0.60
Sliač	−0.08	0.20	0.40	0.49	0.79	0.64	0.12
Boľkovce	0.08	0.18	0.66	0.36	0.47	0.33	0.12
Rimavská Sobota	0.18	0.35	0.51	0.39	0.65	0.54	0.26
Telgárt	−0.21	−0.20	0.32	0.66	0.75	0.60	0.29
Poprad	−0.06	−0.47	0.32	0.46	0.51	0.45	0.28
Košice	−0.25	0.16	0.49	0.51	0.78	0.50	0.15
Milhostov	−0.01	0.19	0.66	0.68	0.81	0.42	−0.02
Kamenica n. Cirochou	−0.38	0.44	0.67	0.31	0.56	0.00	−0.22

Table 4.2 Coefficient of determination (r^2) between vegetation vigor (NDVI) and drought (PDSI) on the selected sites from April to October during the period 2000–2014

Sites (stations)	Months						
	IV.	V.	VI.	VII.	VIII.	IX.	X.
Bratislava	0.06	0.10	0.45	0.39	0.30	0.06	0.06
Piešťany	0.11	0.03	0.22	0.24	0.40	0.42	0.01
Hurbanovo	0.28	0.04	0.08	0.27	0.38	0.25	0.00
Čadca	0.18	0.15	0.27	0.05	0.23	0.54	0.36
Sliač	0.01	0.04	0.16	0.24	0.63	0.40	0.02
Boľkovce	0.01	0.03	0.44	0.13	0.22	0.11	0.01
Rimavská Sobota	0.03	0.12	0.26	0.15	0.42	0.29	0.07
Telgárt	0.04	0.04	0.10	0.44	0.57	0.36	0.08
Poprad	0.00	0.22	0.10	0.21	0.26	0.20	0.08
Košice	0.06	0.03	0.24	0.26	0.61	0.25	0.02
Milhostov	0.00	0.04	0.44	0.46	0.65	0.18	0.00
Kamenica n. Cirochou	0.15	0.19	0.45	0.10	0.31	0.00	0.05

Based on the coefficient of determination (Table 4.2), which expresses share how a change of PDSI affects NDVI, we can state that there were seen high ($0.5 \leq r^2 < 0.8$; highlight with red colour) and moderately ($0.25 \leq r^2 < 0.5$; highlight with green colour) levels of explained variability. These numbers were observed mainly in June (50% of the sites), July (42% of the sites), August (83% of the sites) and in September (50% of the sites).

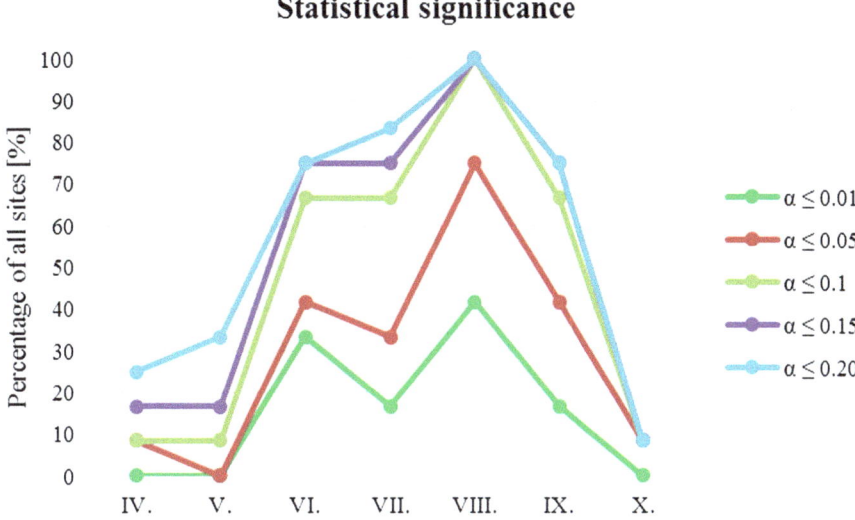

Fig. 4.1 Comparison of p-values at different significance levels on the sites from April to October during the period 2000–2014

P-values were evaluated at five significance levels: 0.01, 0.05, 0.1, 0.15 and 0.2 (Fig. 4.1). There was not observed 100% share of significant p-values on all monitored sites in any month at significance levels 0.01 and 0.05. Statistical significance on all sites was identified in August at significance level 0.1. Statistical significance at significance level 0.1 was proved on 8% of the sites in April, 8% of the sites in May, 67% of the sites in June, 67% of the sites in July, 67% of the sites in September and 8% of sites in October. If we reduced the statistical significance at levels 0.15 or 0.2, it would not bring a substantial change in the percentage of significant p-values on the sites. Share of the sites in August and October remained identical at the last three levels of significance. Share of the sites with significant p-values in other months only slightly increased. It means it is not necessary to decrease under significance level 0.1.

The results show that the strongest relationships between NDVI and PDSI were observed from June to August. Weaker correlations, so less impact of drought on vegetation vigor on arable land, were recorded at the beginning (April and May) and at the end (September and October) of the growing season. It is clear, that reaction of vegetation to drought varies in each month. So, the effect of drought on vegetation depends on phenophase in which crops are. It is also possible to say, that the increasing linear trends were recorded on all sites in months with the strongest correlations (June–September). It means that the increasing values of PDSI led to increasing values of NDVI. Contrariwise, when the PDSI values were low, the NDVI values were low, too. Thus, the impact of drought was reflected in the worse state of vegetation. Findings indicate that the NDVI is useful for assessing the drought in conditions of Slovakia.

4.1.1   Impacts on Crop Yield Forecasts

The relationship between detrended yields of selected crops in kg.ha^{-1} and monthly NDVI on a district level during the period of 2000–2014 was expressed by a linear function. For these correlations, Pearson's correlation coefficient and coefficient of determination were calculated. As statistically significant correlations were considered these with p-values $\alpha \leq 0.05$ and $\alpha \leq 0.01$.

The results show that the impact of drought on spring barley yields was observed mainly in the Podunajská nížina Valley, respectively in the production areas of western Slovakia. Statistical significance of correlations at the significance level $\alpha \leq 0.05$ was demonstrated in 28 districts (42% of districts) of 67 evaluated (Figs. 4.2, 4.3, 4.4, 4.5, 4.6, 4.7 and 4.8). The analysis of relationships in individual months showed that the spring barley yields were affected by drought especially in April, when significant correlations were recorded in 17 districts (61% of the evaluated districts)

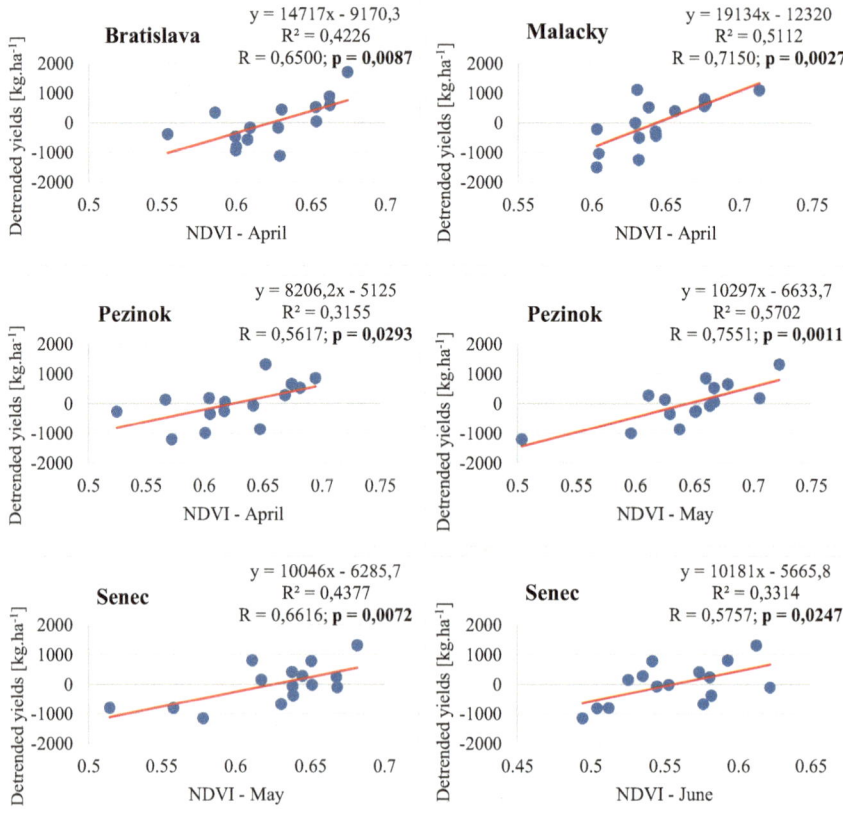

Fig. 4.2 Statistically significant correlations between detrended spring barley yields in kg.ha^{-1} and monthly NDVI values in individual districts in Bratislava region (2000–2014)

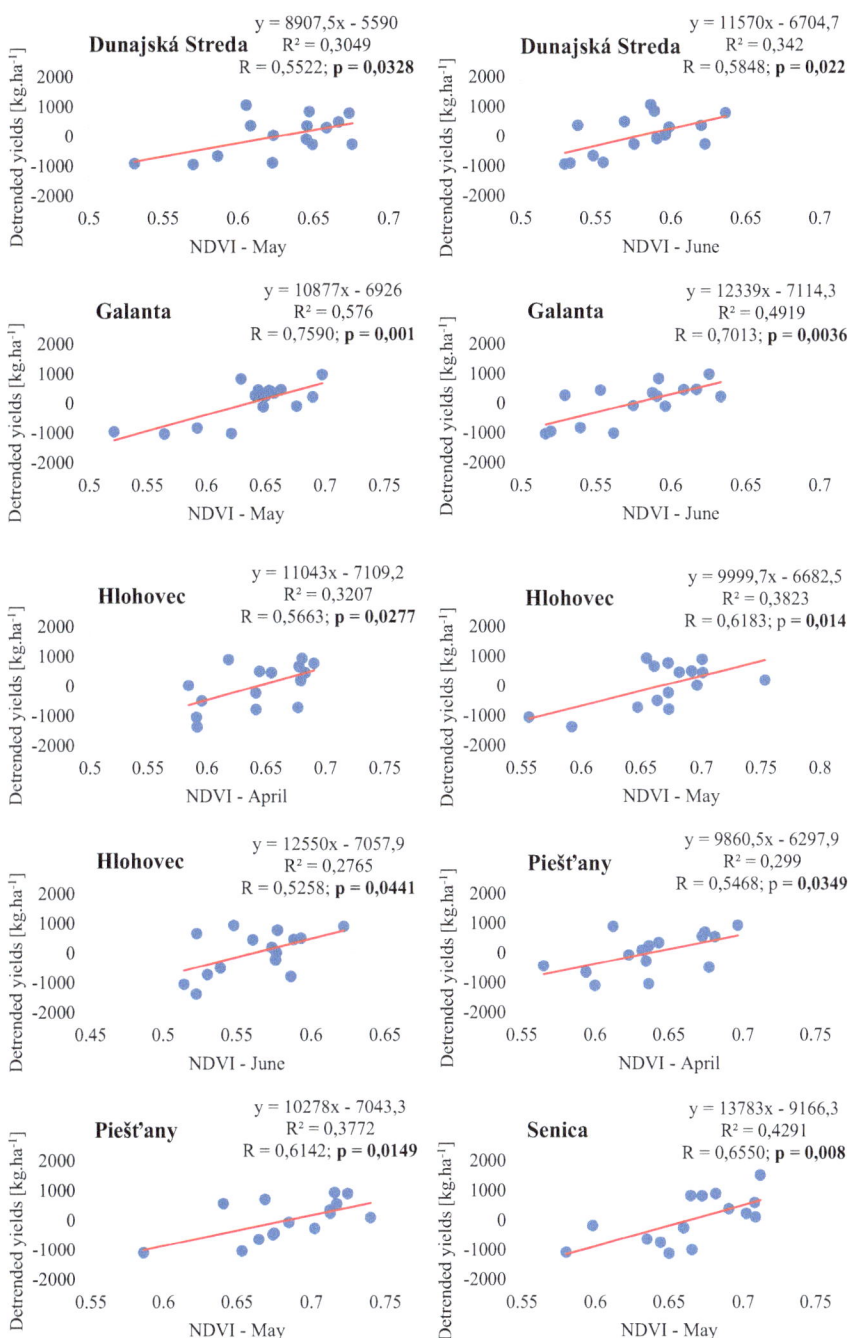

Fig. 4.3 Statistically significant correlations between detrended spring barley yields in kg.ha^{-1} and monthly NDVI values in individual districts in Trnava region (2000–2014)

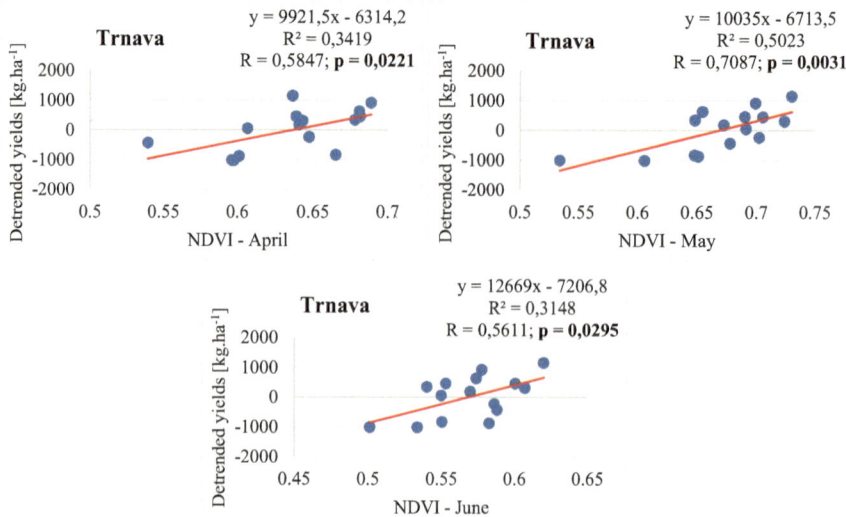

Fig. 4.3 (continued)

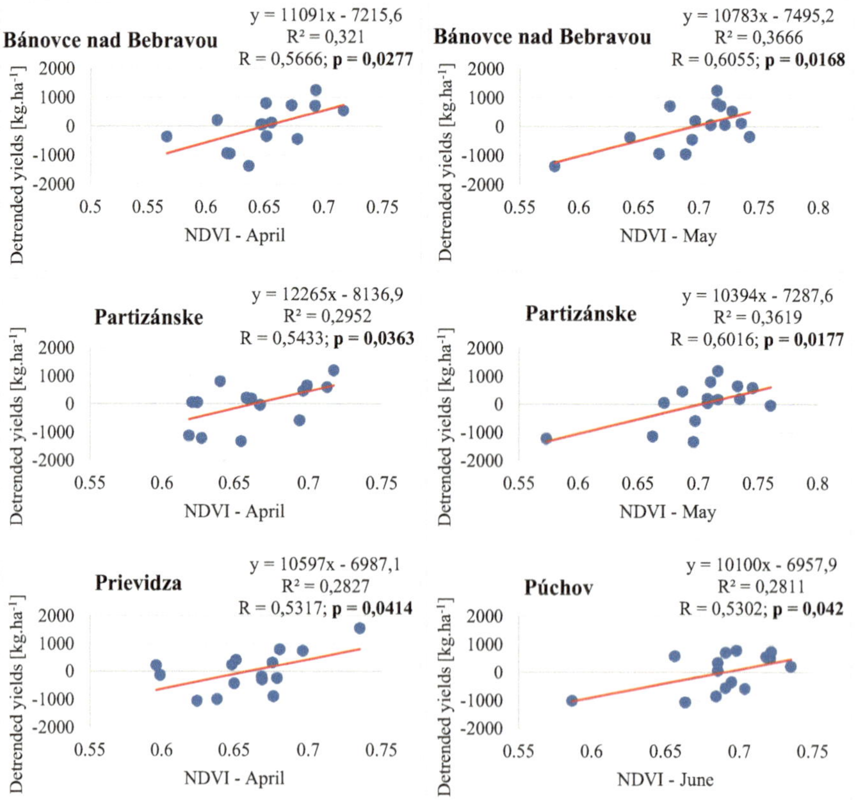

Bánovce nad Bebravou
y = 11091x − 7215,6
R² = 0,321
R = 0,5666; **p = 0,0277**
NDVI - April

Bánovce nad Bebravou
y = 10783x − 7495,2
R² = 0,3666
R = 0,6055; **p = 0,0168**
NDVI - May

Partizánske
y = 12265x − 8136,9
R² = 0,2952
R = 0,5433; **p = 0,0363**
NDVI - April

Partizánske
y = 10394x − 7287,6
R² = 0,3619
R = 0,6016; **p = 0,0177**
NDVI - May

Prievidza
y = 10597x − 6987,1
R² = 0,2827
R = 0,5317; **p = 0,0414**
NDVI - April

Púchov
y = 10100x − 6957,9
R² = 0,2811
R = 0,5302; **p = 0,042**
NDVI - June

Fig. 4.4 Statistically significant correlations between detrended spring barley yields in kg.ha⁻¹ and monthly NDVI values in individual districts in Trenčín region (2000–2014)

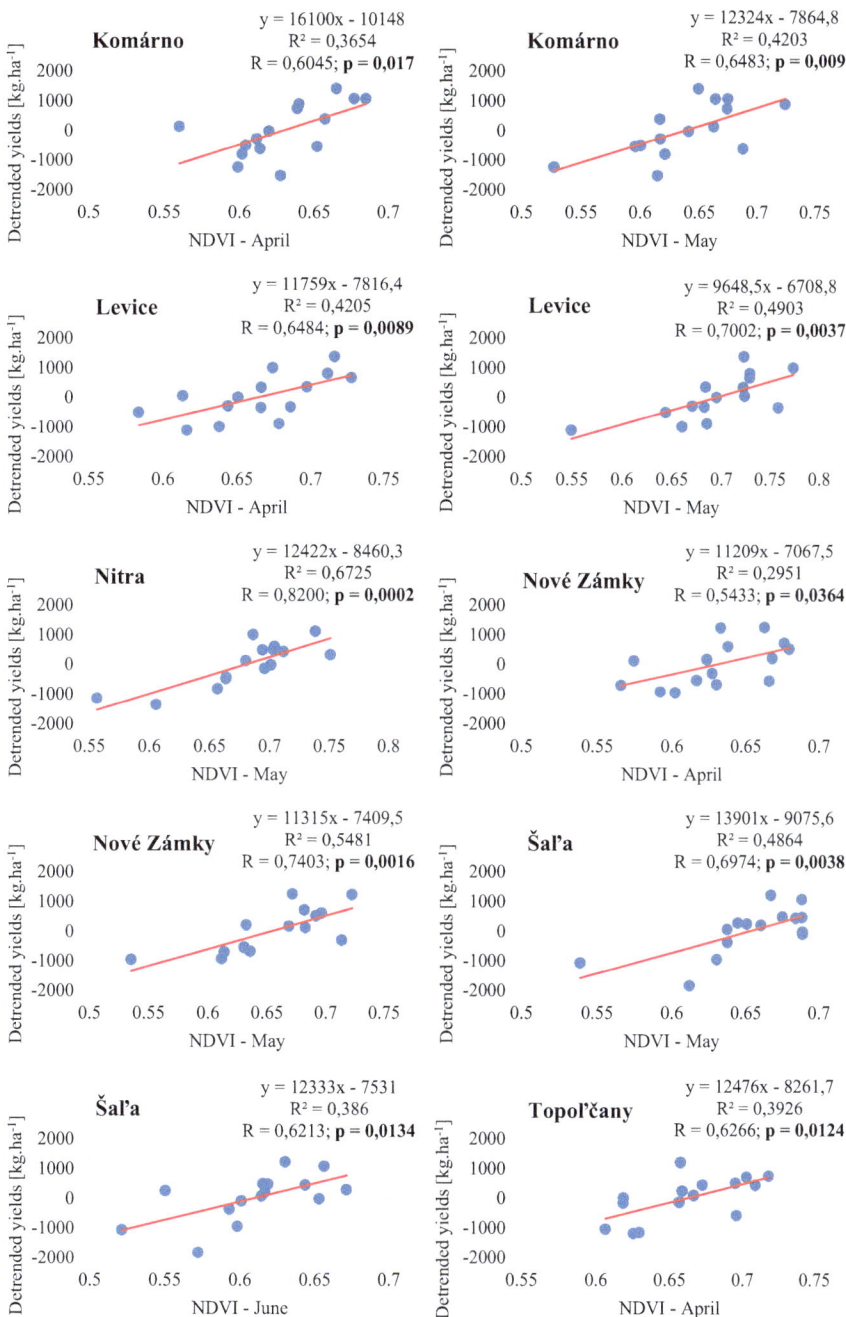

Fig. 4.5 Statistically significant correlations between detrended spring barley yields in kg.ha⁻¹ and monthly NDVI values in individual districts in Nitra region (2000–2014)

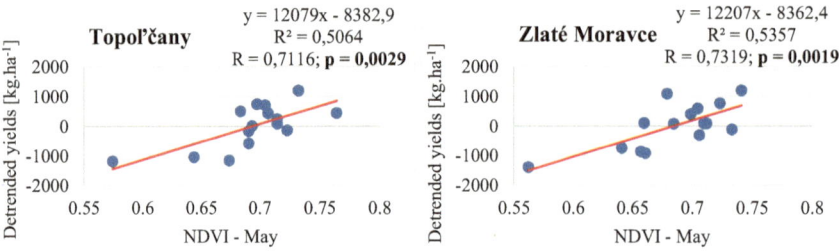

Fig. 4.5 (continued)

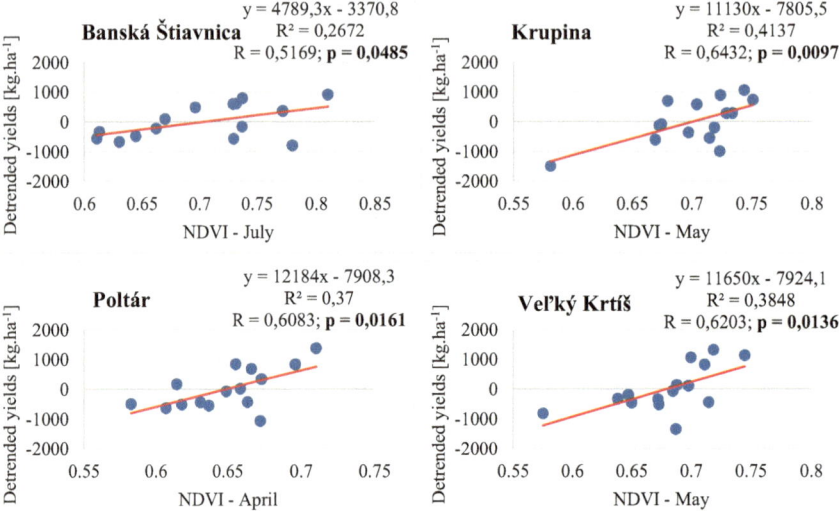

Fig. 4.6 Statistically significant correlations between detrended spring barley yields in kg.ha⁻¹ and monthly NDVI values in individual districts in Banská Bystrica region (2000–2014)

and May with significant correlations in 19 districts, it means 68% of the districts, where the statistically significant impact of drought on the yields was detected. Drought affected spring barley yields in 7 districts in June (25% of the evaluated districts) and in Banská Štiavnica district in July (4% of the evaluated districts). At the level of significance $\alpha \leq 0.05$, there was statistically least significant correlation in Banská Štiavnica district in July with a correlation coefficient of 0.52 and a coefficient of determination of 0.27.

Statistically significant correlations at the significance level $\alpha \leq 0.01$ between detrended spring barley yields and monthly NDVI were identified in 15 districts

Fig. 4.7 Statistically significant correlations between detrended spring barley yields in kg.ha⁻¹ and monthly NDVI values in individual districts in Prešov region (2000–2014)

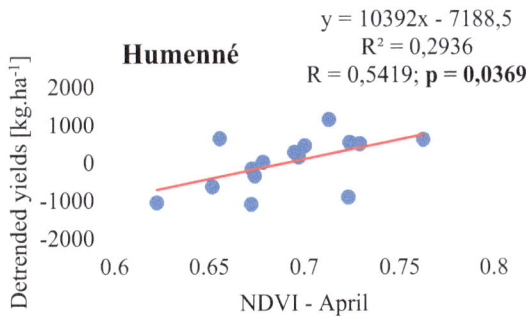

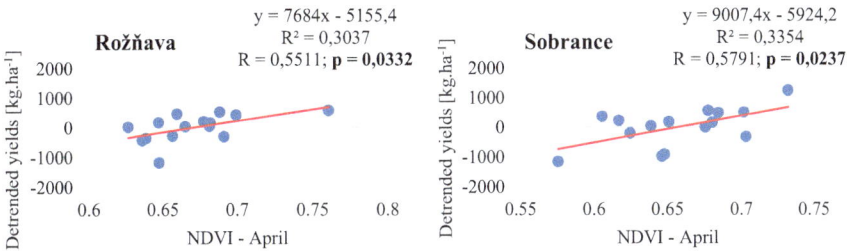

Fig. 4.8 Statistically significant correlations between detrended spring barley yields in kg.ha⁻¹ and monthly NDVI values in individual districts in Košice region (2000–2014)

(22% from all evaluated districts). Analysis of the individual months showed that spring barley yields were influenced by drought in three districts in April (20% of the evaluated districts). The most substantial influence of drought on yields was proved in May specifically in 13 districts (87%). Statistically significant correlation in June was seen only in one district (7% of the evaluated districts). Impact of drought in July was not observed.

The most significant correlation according to p-value was detected in Nitra district in May with the correlation coefficient 0.82 and the coefficient of determination 0.67. Contrariwise, the least significant correlation was observed in Krupina district in May with a correlation coefficient of 0.64 and a coefficient of determination 0.41. It represents only one significant district of whole Banská Bystrica region.

Statistically significant correlations in each district are shown in maps for individual months (Figs. 4.9, 4.10 and 4.11).

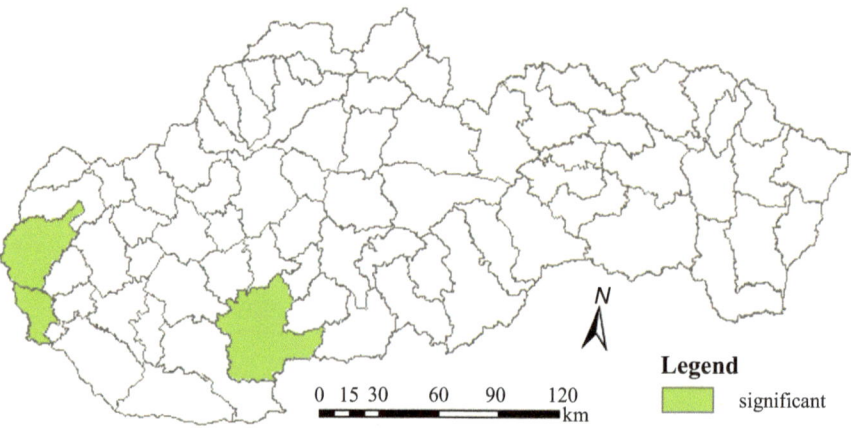

Fig. 4.9 Statistically significant (p ≤ 0.01) correlations between detrended yields of spring barley [kg.ha⁻¹] and NDVI values on a district level in April during the period 2000–2014

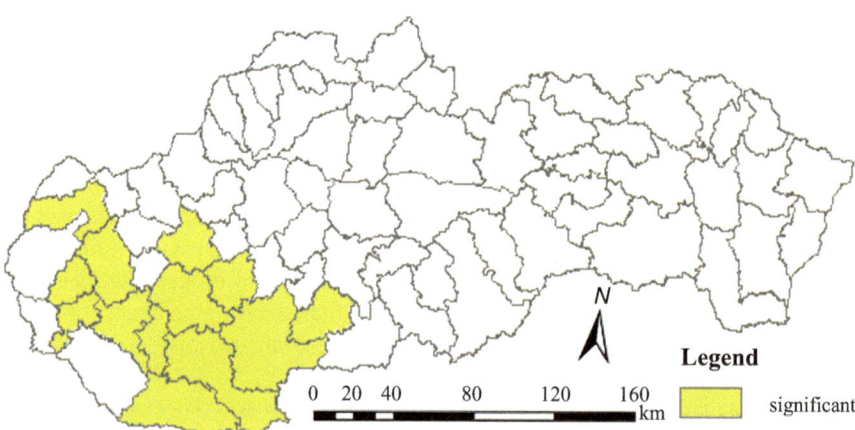

Fig. 4.10 Statistically significant (p ≤ 0.01) correlations between detrended yields of spring barley [kg.ha⁻¹] and NDVI values on a district level in May during the period 2000–2014

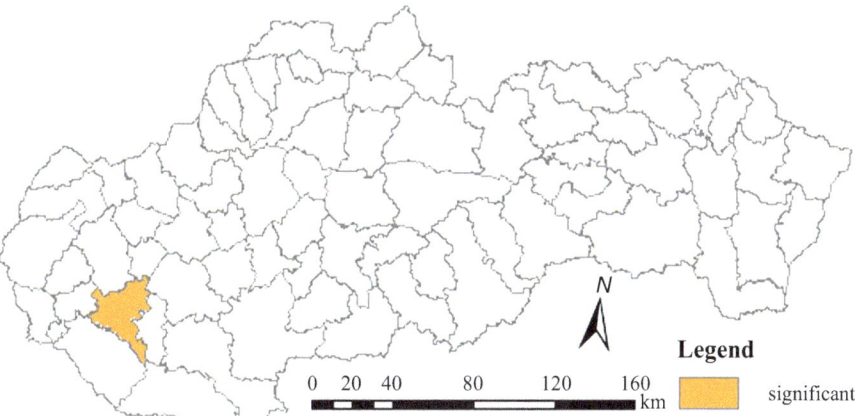

Fig. 4.11   Statistically significant (p ≤ 0.01) correlations between detrended yields of spring barley [kg.ha^{-1}] and NDVI values on a district level in June during the period 2000–2014

Statistically significant correlations on the level of significance $\alpha \leq 0.05$ in case of winter wheat during its growing season were identified in 60 districts of 70 evaluated (Figs. 4.12, 4.13, 4.14, 4.15, 4.16, 4.17, 4.18 and 4.19). This represented a share of 86%. Analysis of the individual months showed that impact of drought on final winter wheat yields was detected in 12 districts in January (20% of the evaluated districts), in 16 districts in February (27% of the evaluated districts) and in 42 districts in March (70% of the evaluated districts). The most substantial impact of drought was observed in April in 46 districts (77% of the evaluated districts). Drought affected yields in 16 districts in May (27% of the evaluated districts), in 8 districts in June (13% of the evaluated districts), in 7 districts in October (12% of the evaluated districts), in 14 districts in November (23% of the evaluated districts) and in 2 districts in December (3% of the evaluated districts). However the significant correlations in winter months are illogical, because of growing season of winter wheat. The least significant correlation at the significance level $\alpha \leq 0.05$ was observed in the Malacky district in April with a correlation coefficient of 0.52 and a coefficient of determination of 0.27.

At the level of statistical significance $\alpha \leq 0.01$ significant correlations were observed between winter wheat yields and NDVI in 27 districts, which was 39% of all evaluated districts. Drought affected yields in 3 districts in January (11% of the evaluated districts), in 11 districts in March (41% of the evaluated districts). The most intense drought impact was identified in 18 districts in April (67% of the

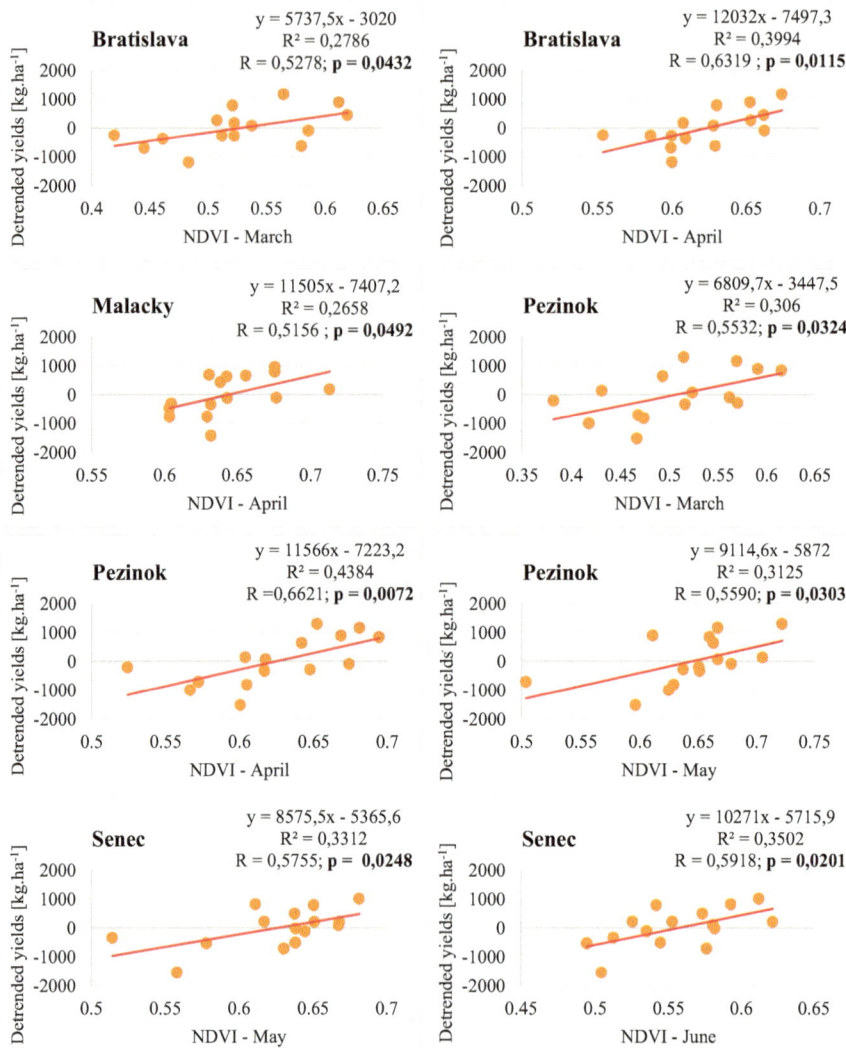

Fig. 4.12 Statistically significant correlations between detrended winter wheat yields in kg.ha⁻¹ and monthly NDVI values in individual districts in Bratislava region (2000–2014)

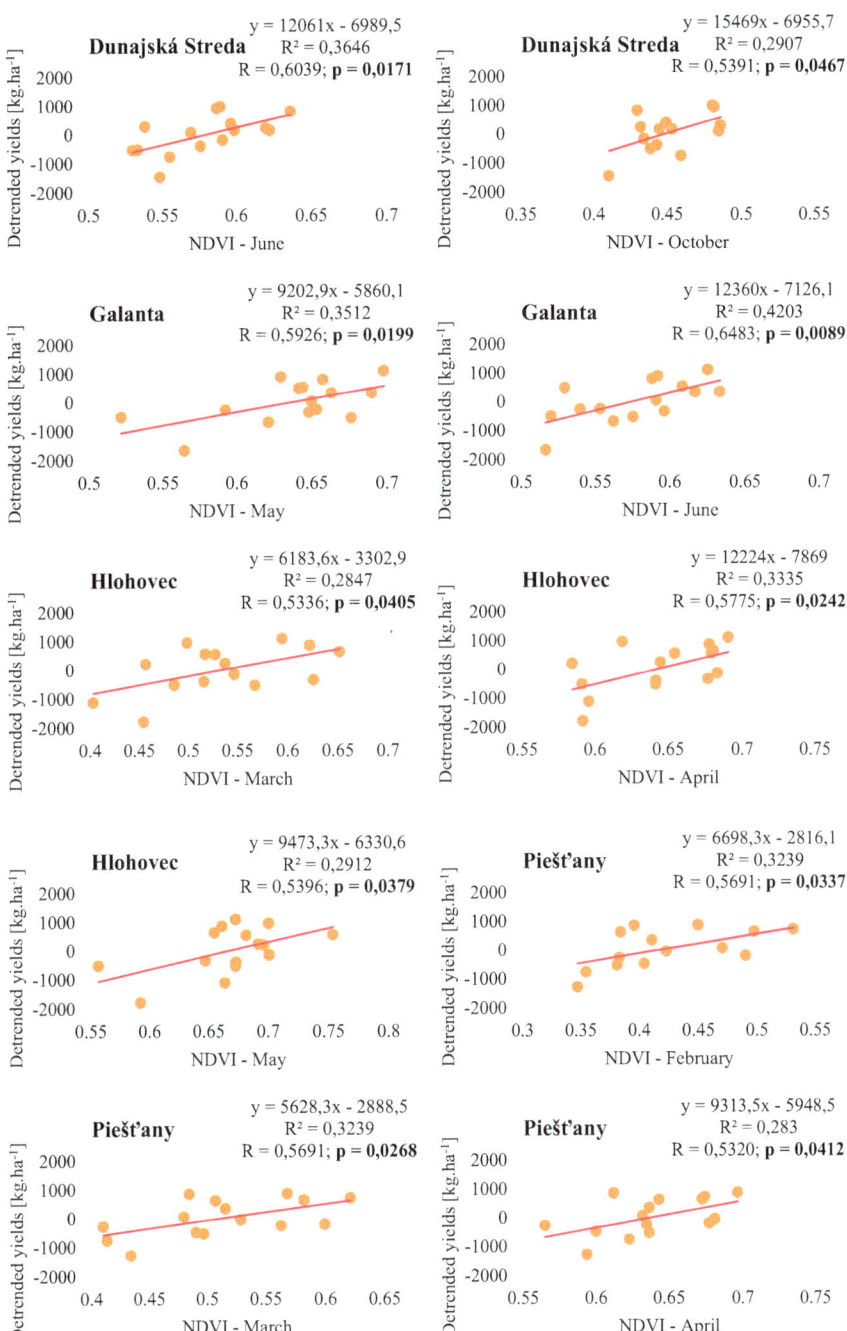

Fig. 4.13 Statistically significant correlations between detrended winter wheat yields in kg.ha^{-1} and monthly NDVI values in individual districts in Trnava region (2000–2014)

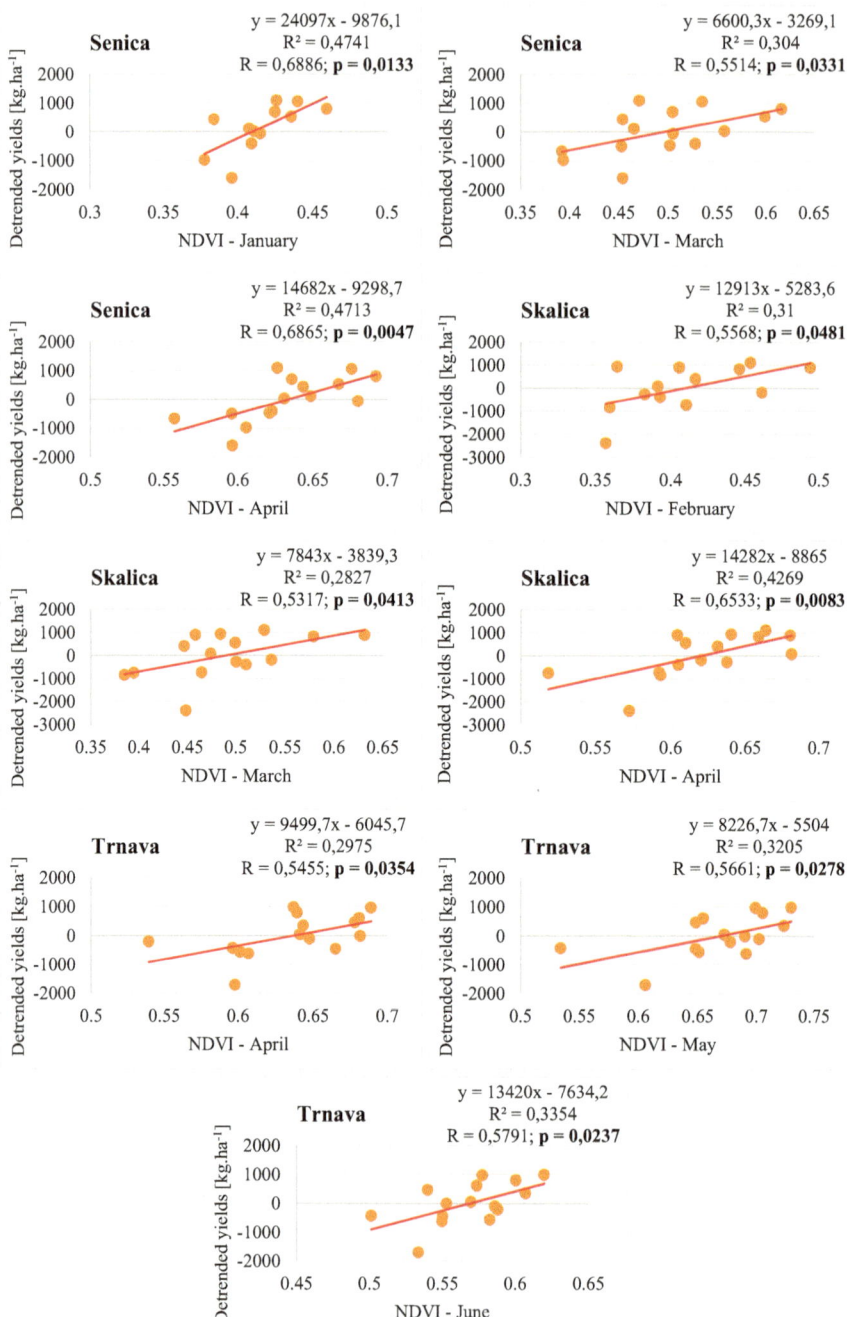

Fig. 4.13 (continued)

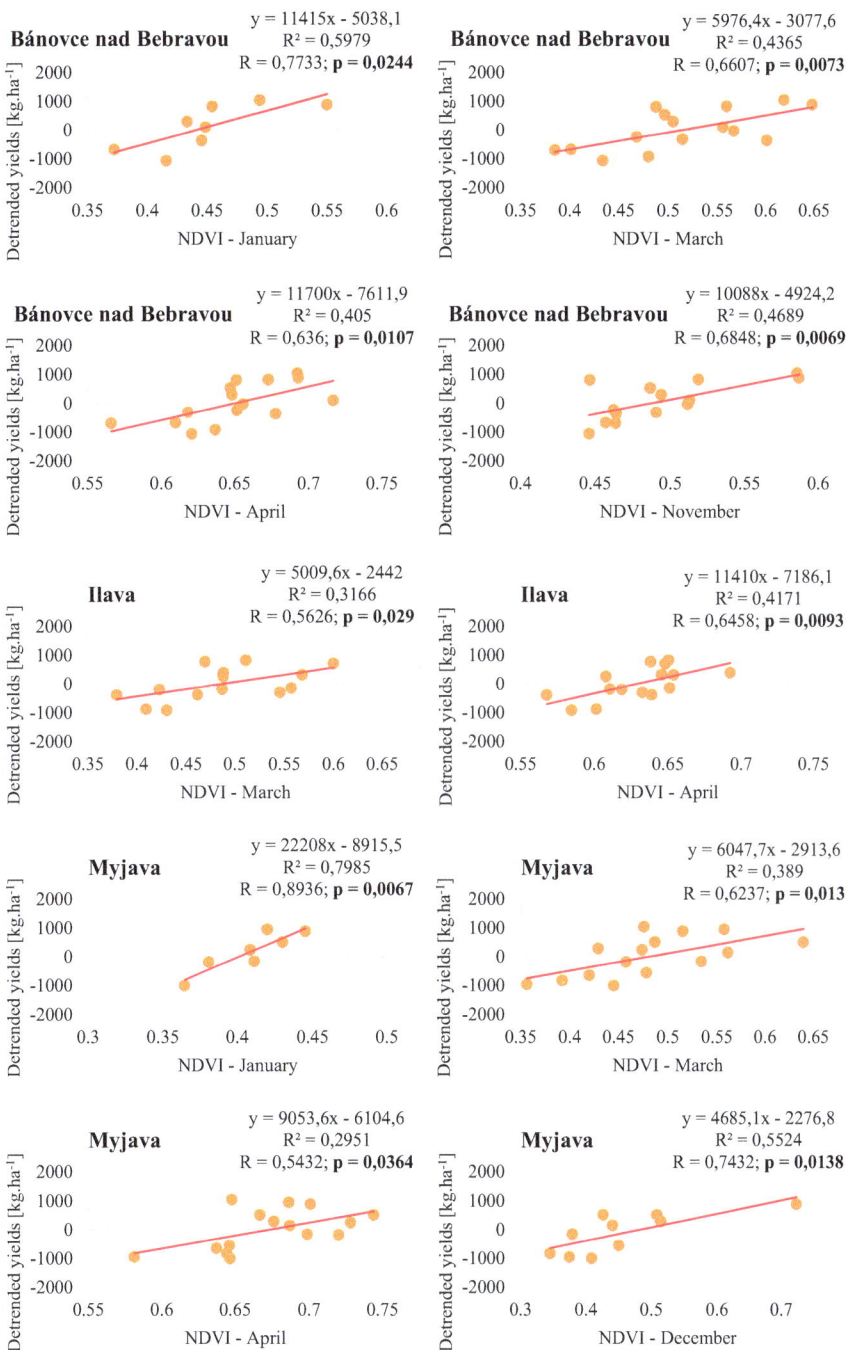

Fig. 4.14 Statistically significant correlations between detrended winter wheat yields in kg.ha⁻¹ and monthly NDVI values in individual districts in Trenčín region (2000–2014)

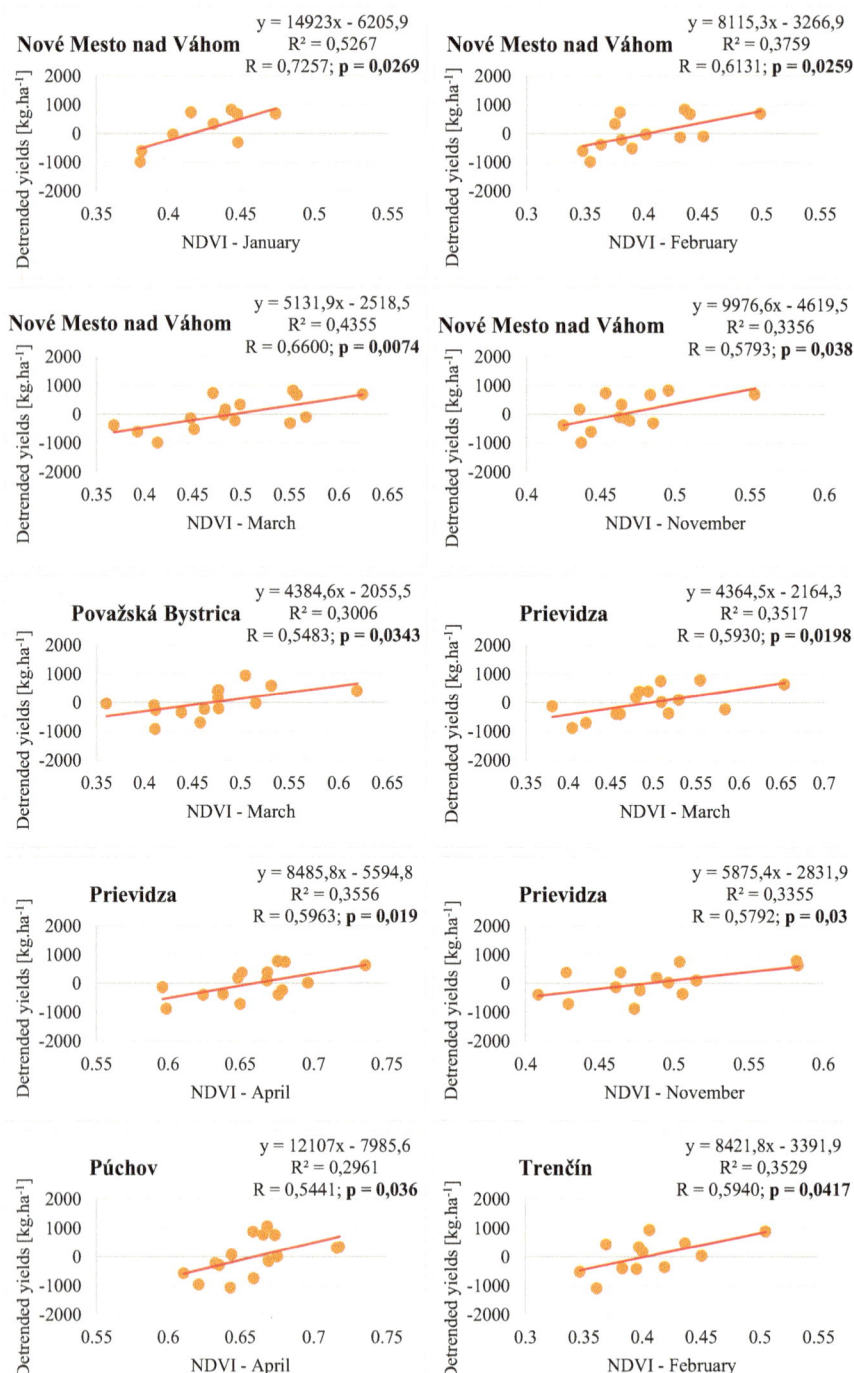

Fig. 4.14   (continued)

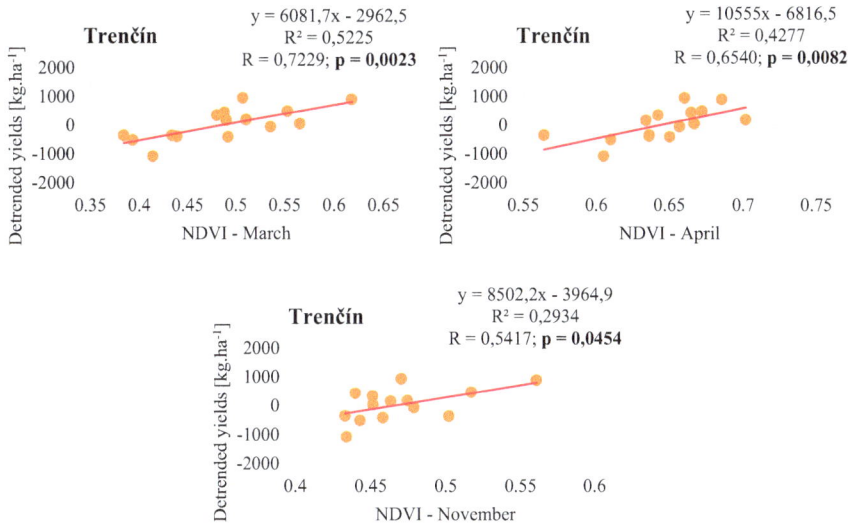

Fig. 4.14 (continued)

evaluated districts). Drought in May affected the yields in part of the Podunajská nížina Valley in five districts (19% of the evaluated districts). Impact of the drought was observed in Galanta district in June (4% of the evaluated districts), in two districts in October (7% of the evaluated districts) and in three districts in November (11% of the evaluated districts). In February and December, there were no observations of significant correlations at this level of statistical significance. The most significant correlation was detected in the districts of Zlaté Moravce in April with $r = 0.78$ and $r^2 = 0.61$. The least significant correlation was observed in two districts, the Liptovský Mikuláš district in April with $r = 0.64$ and $r^2 = 0.41$ and the Levice district with $r = 0.66$ and $r^2 = 0.44$.

Map of significant correlations at significance level $\alpha \leq 0.01$ between winter wheat yields and NDVI at the district level in an individual month (January, March, April, May, June, October and November) is shown in Figs. 4.20, 4.21, 4.22, 4.23, 4.24, 4.25 and 4.26.

At the level of significance $\alpha \leq 0.05$, statistically significant correlations between NDVI and detrended maize yields were noticed in 18 districts (44%) of the 41 districts that were evaluated (Figs. 4.27, 4.28, 4.29, 4.30 and 4.31). Based on an analysis of the relationships in individual months of the growing season, the impact of drought on the maize yields was most significant in July, when significant correlations were detected in 16 districts (89% of the evaluated districts). Drought affected yields in four districts in June (22% of the evaluated districts) and only in Šaľa district in May (6% of the evaluated districts). No significant correlations between drought and yields were found in other months of the growing season of maize. The

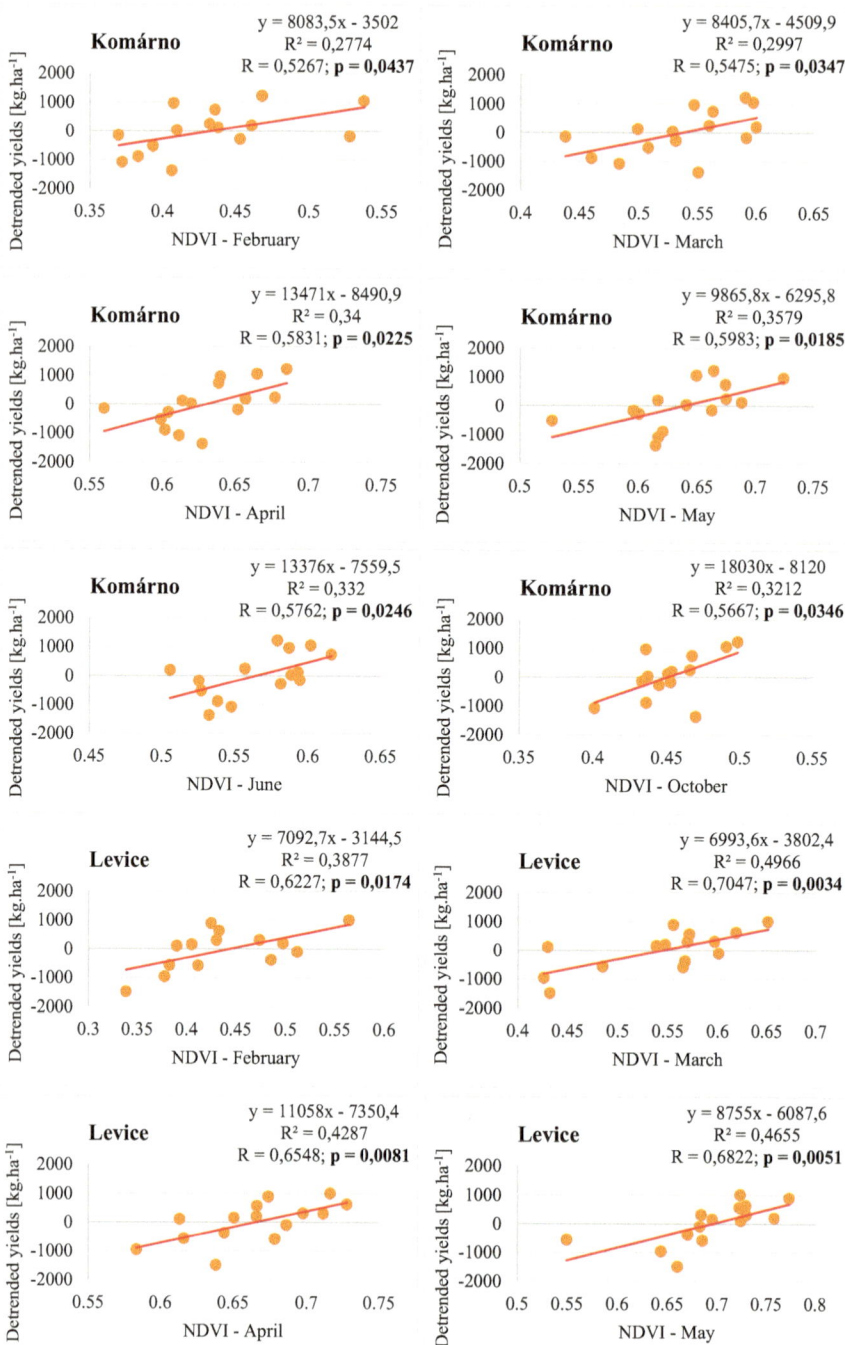

Fig. 4.15 Statistically significant correlations between detrended winter wheat yields in kg.ha^{-1} and monthly NDVI values in individual districts in Nitra region (2000–2014)

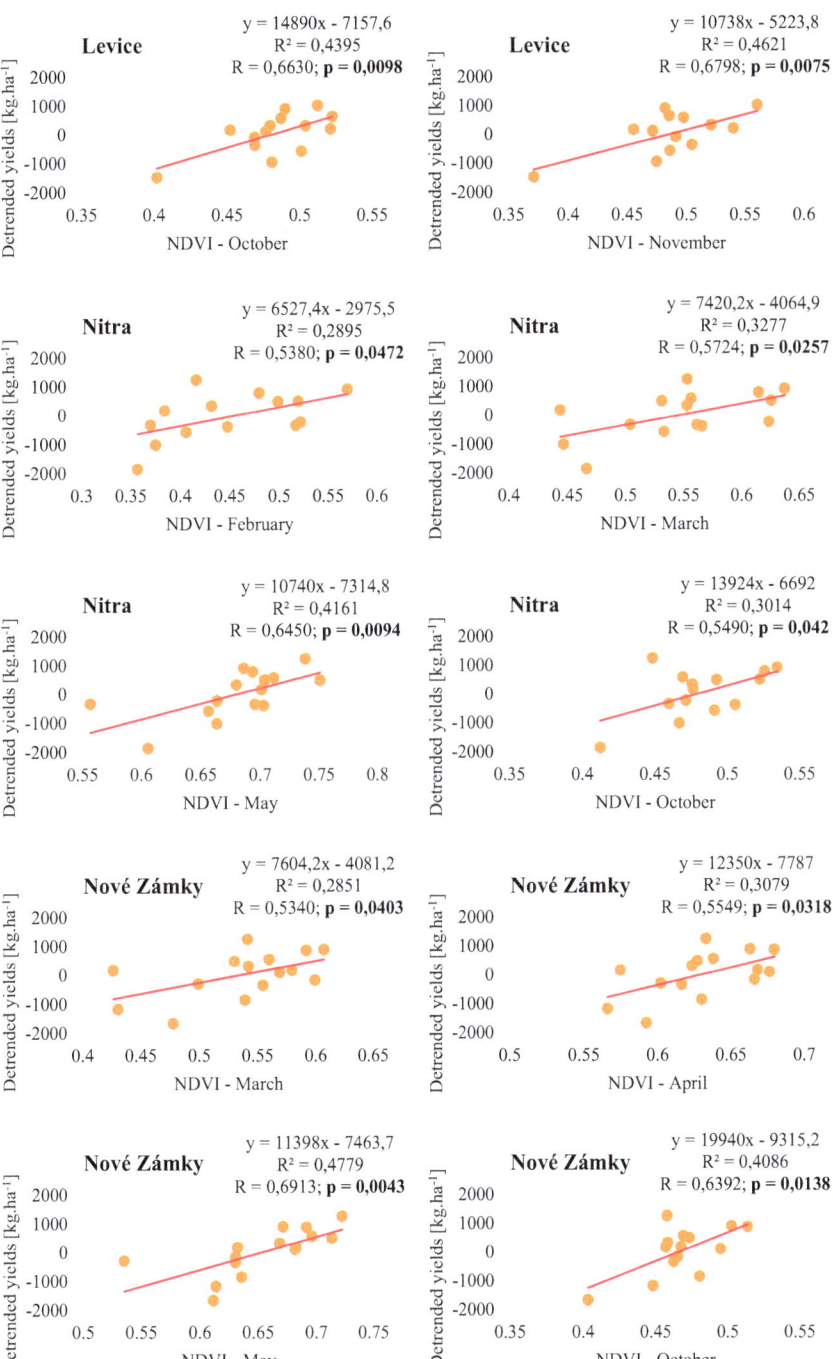

Fig. 4.15 (continued)

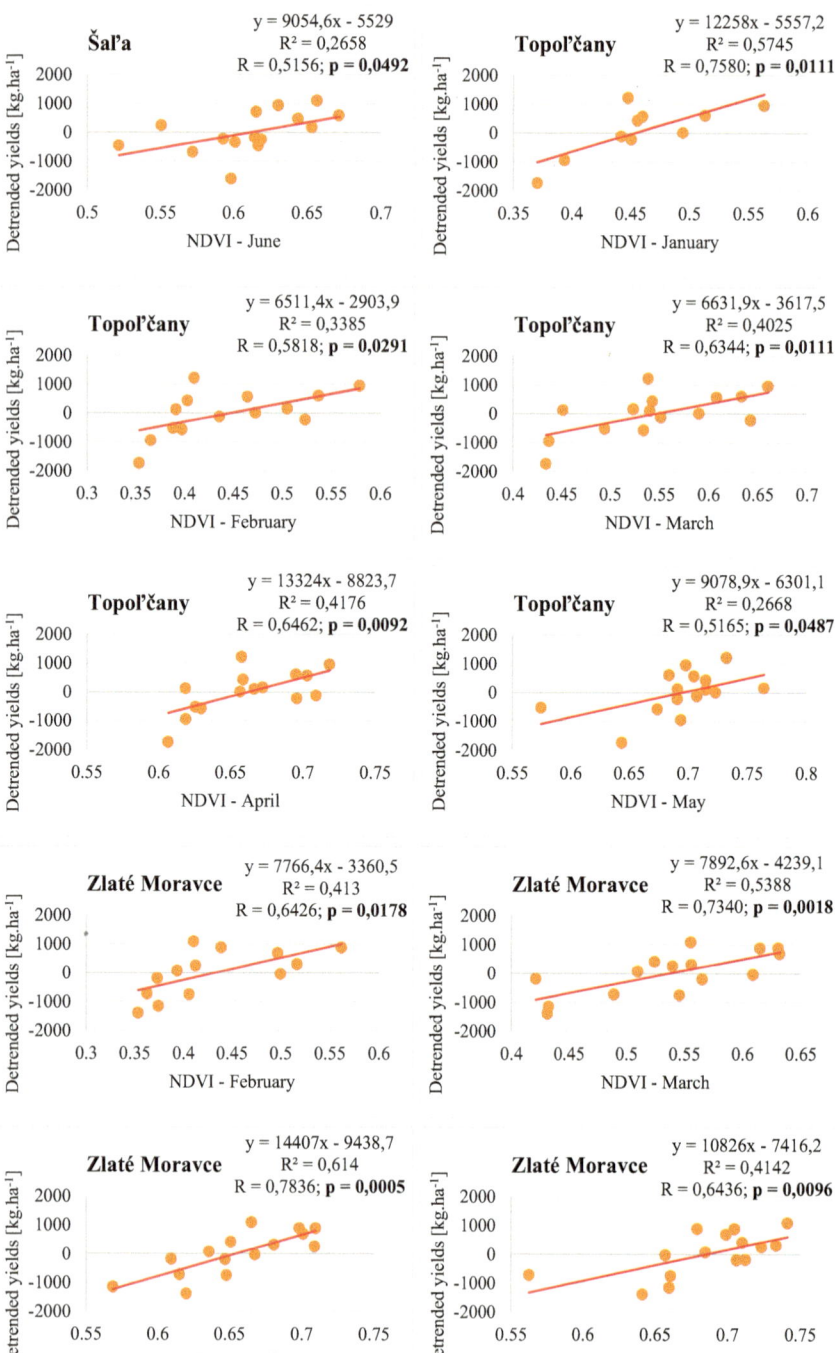

Fig. 4.15 (continued)

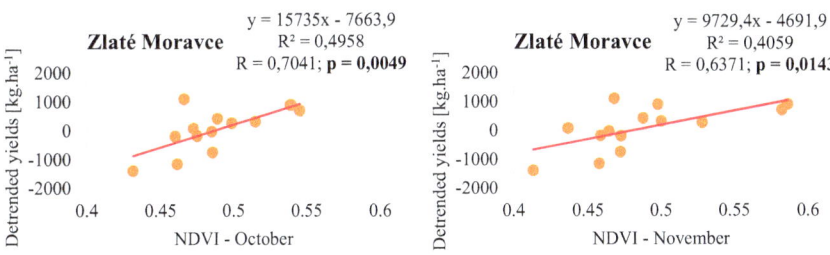

Zlaté Moravce
y = 15735x − 7663,9
R² = 0,4958
R = 0,7041; **p = 0,0049**

Detrended yields [kg.ha⁻¹]
NDVI - October

Zlaté Moravce
y = 9729,4x − 4691,9
R² = 0,4059
R = 0,6371; **p = 0,0143**

Detrended yields [kg.ha⁻¹]
NDVI - November

Fig. .15 (continued)

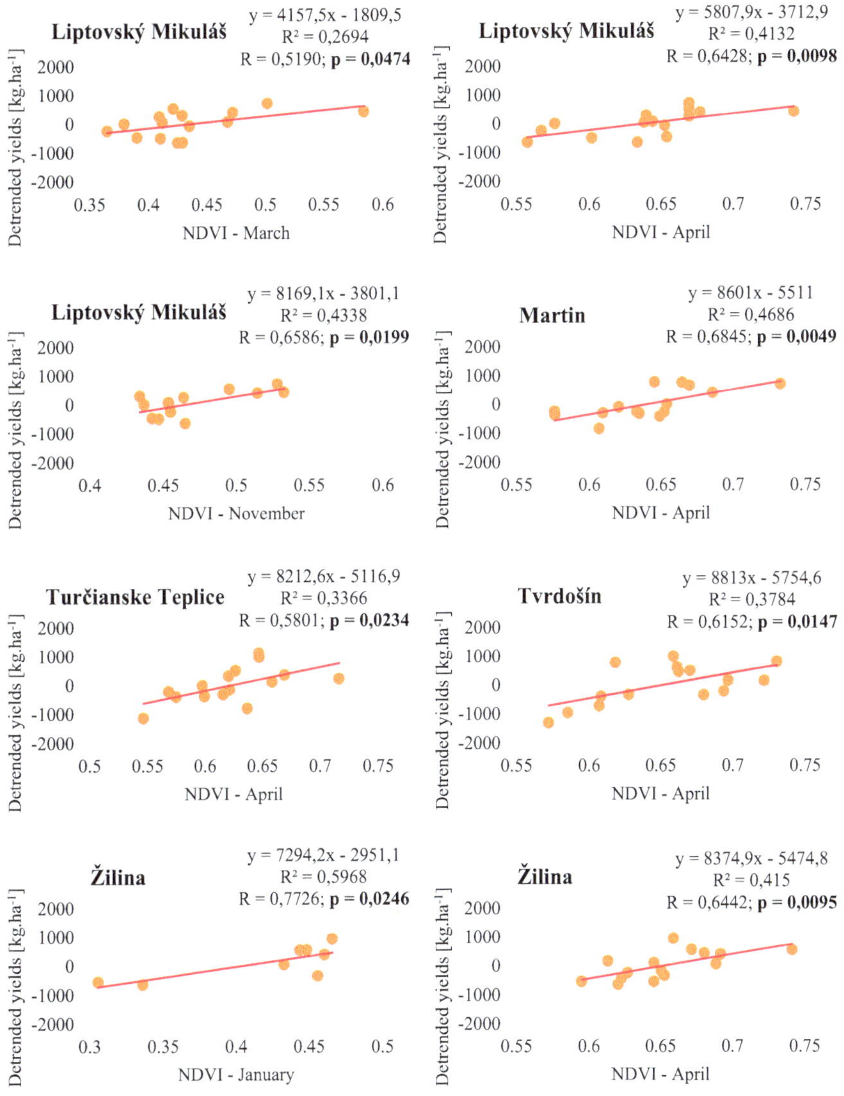

Liptovský Mikuláš
y = 4157,5x − 1809,5
R² = 0,2694
R = 0,5190; **p = 0,0474**

Detrended yields [kg.ha⁻¹]
NDVI - March

Liptovský Mikuláš
y = 5807,9x − 3712,9
R² = 0,4132
R = 0,6428; **p = 0,0098**

Detrended yields [kg.ha⁻¹]
NDVI - April

Liptovský Mikuláš
y = 8169,1x − 3801,1
R² = 0,4338
R = 0,6586; **p = 0,0199**

Detrended yields [kg.ha⁻¹]
NDVI - November

Martin
y = 8601x − 5511
R² = 0,4686
R = 0,6845; **p = 0,0049**

Detrended yields [kg.ha⁻¹]
NDVI - April

Turčianske Teplice
y = 8212,6x − 5116,9
R² = 0,3366
R = 0,5801; **p = 0,0234**

Detrended yields [kg.ha⁻¹]
NDVI - April

Tvrdošín
y = 8813x − 5754,6
R² = 0,3784
R = 0,6152; **p = 0,0147**

Detrended yields [kg.ha⁻¹]
NDVI - April

Žilina
y = 7294,2x − 2951,1
R² = 0,5968
R = 0,7726; **p = 0,0246**

Detrended yields [kg.ha⁻¹]
NDVI - January

Žilina
y = 8374,9x − 5474,8
R² = 0,415
R = 0,6442; **p = 0,0095**

Detrended yields [kg.ha⁻¹]
NDVI - April

Fig. 4.16 Statistically significant correlations between detrended winter wheat yields in kg.ha⁻¹ and monthly NDVI values in individual districts in Žilina region (2000–2014)

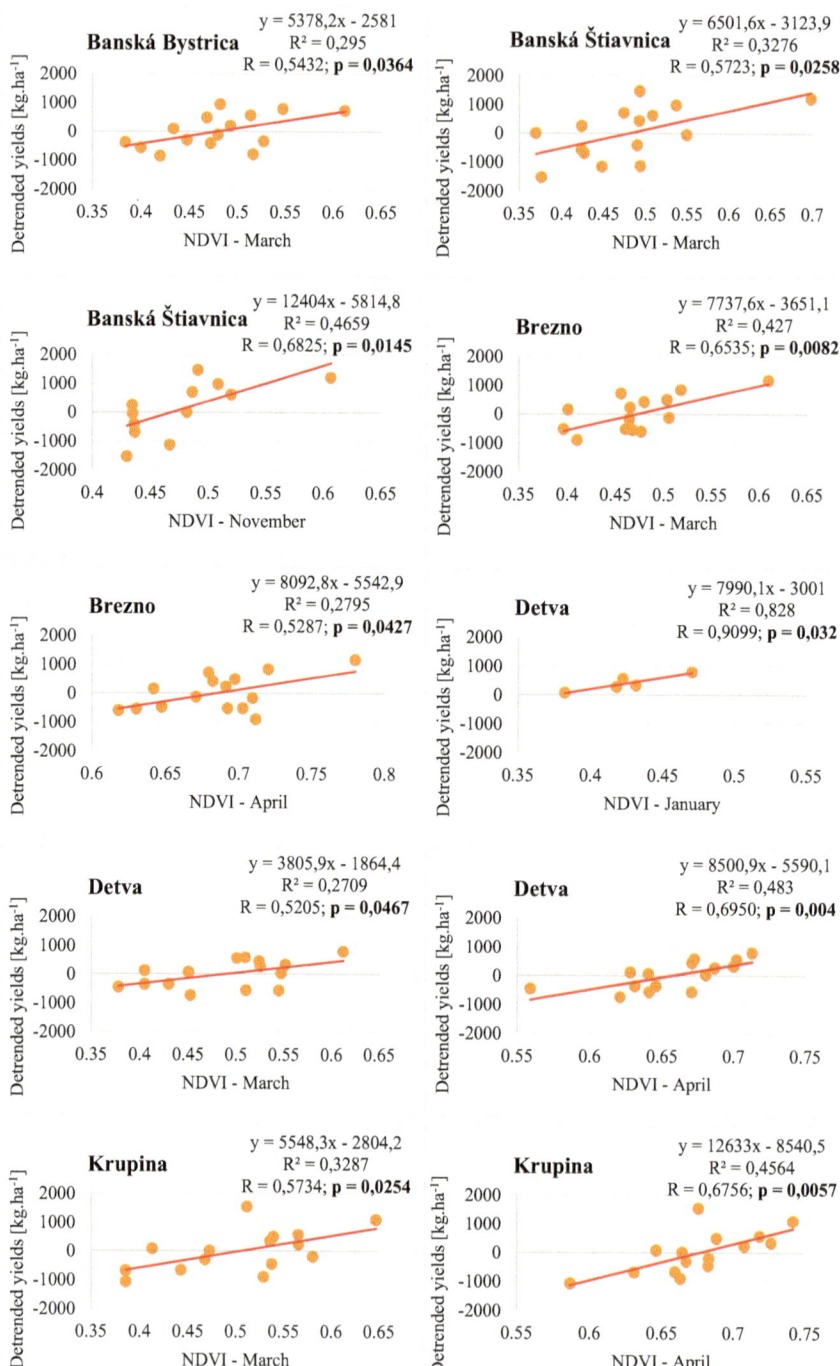

Fig. 4.17 Statistically significant correlations between detrended winter wheat yields in kg.ha⁻¹ and monthly NDVI values in individual districts in Banská Bystrica region (2000–2014)

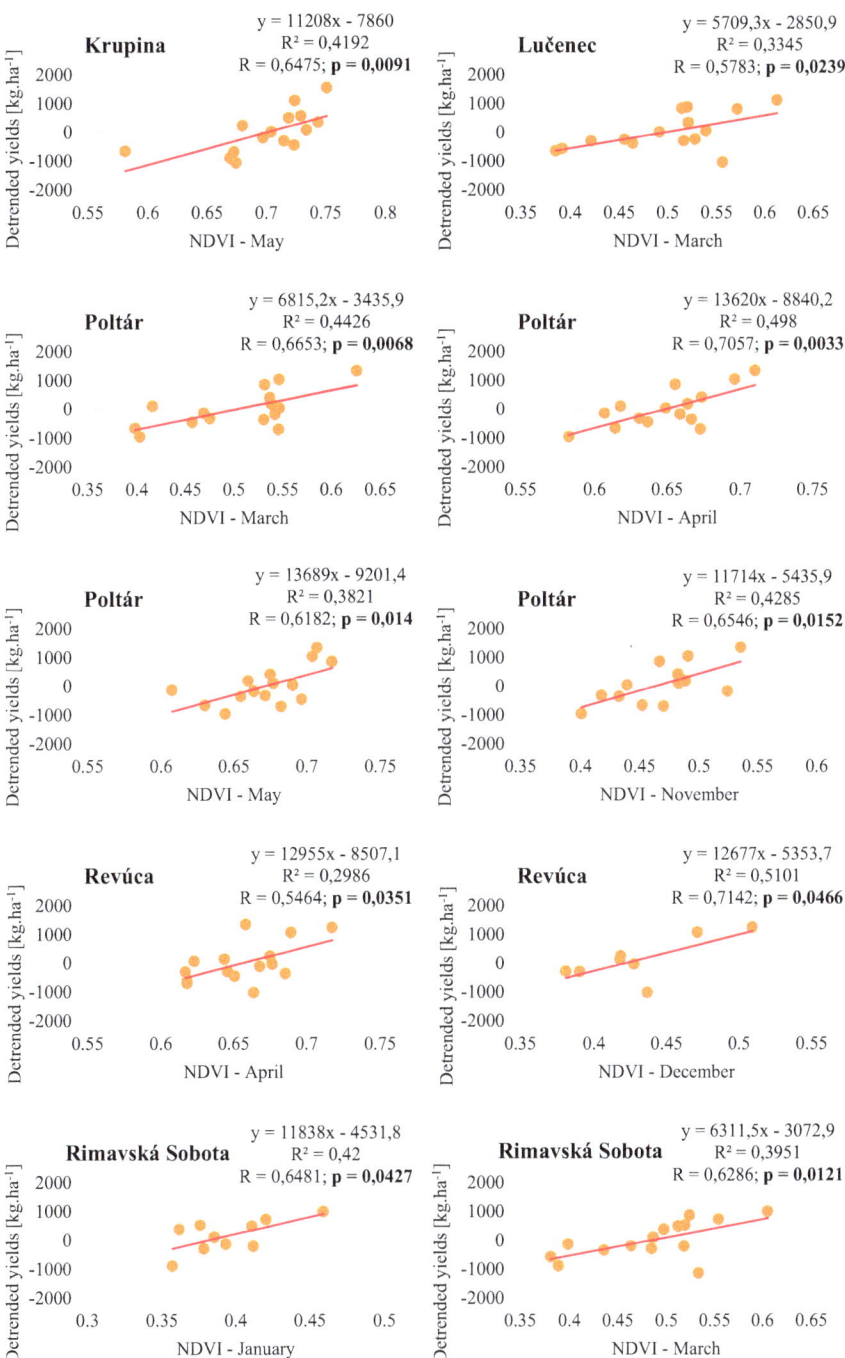

Fig. 4.17   (continued)

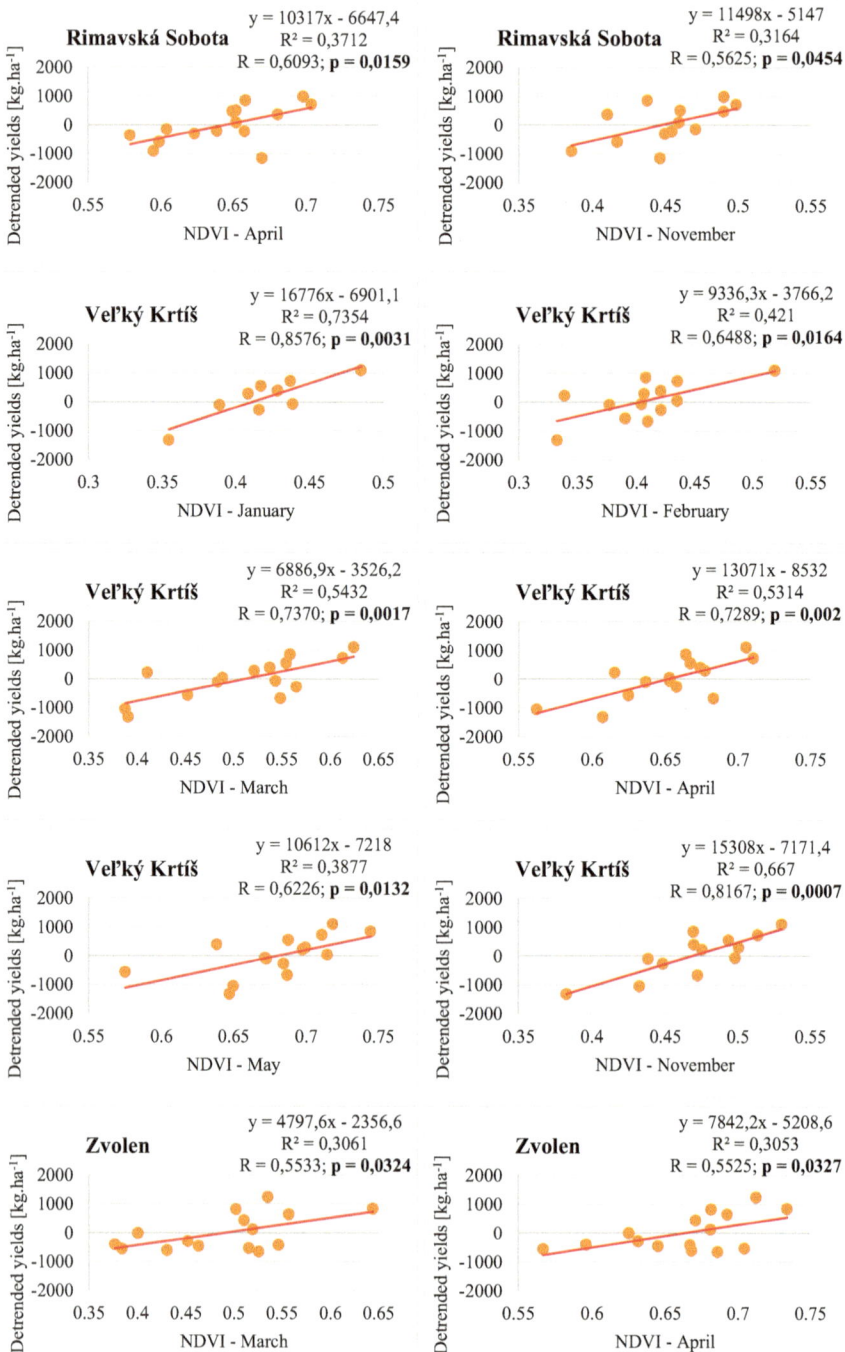

Fig. 4.17 (continued)

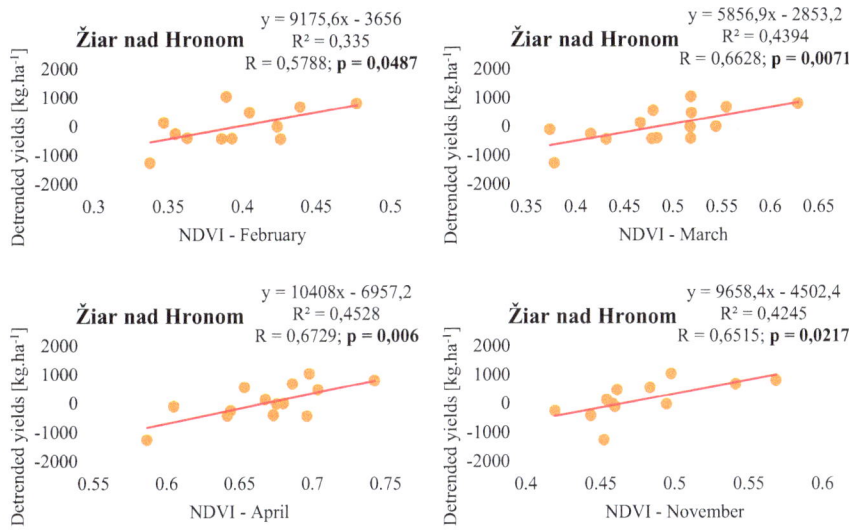

Fig. 4.17 (continued)

least significant correlation between maize yields and NDVI values was recorded in Galanta district in June with a correlation coefficient of 0.52 and a coefficient of determination of 0.27.

At the significance level $\alpha \leq 0.01$, significant correlations were observed only in three districts (7% of all evaluated districts). The monthly analysis confirmed a significant relationship between the maize yields and NDVI in 2 months. In May in the district of Šaľa (33% of the evaluated districts) and in July in two districts (67% of the evaluated districts). The most significant correlation was recorded in the Nové Mesto nad Váhom district in July with a correlation coefficient of 0.72 and a coefficient of determination of 0.51. The least significant correlation was found in the Šaľa district in May with a correlation coefficient of 0.65 and a coefficient of determination of 0.42.

The map of statistically significant correlations ($\alpha \leq 0.01$) between the detrended corn yields and NDVI values at the district level in May and July is shown in the figures (Figs. 4.32 and 4.33).

4.1.2 Practical Use

In this thesis, the applicability of the NDVI for the drought assessment was examined. This method adequately identifies the effect of moisture deficiency on yields of selected field crops. The advantage of remote sensing data is the monitoring the changes in vegetation vigor at the local, regional and global level. It allows time-efficient mapping with exact spatial information about crops without any financial requirement.

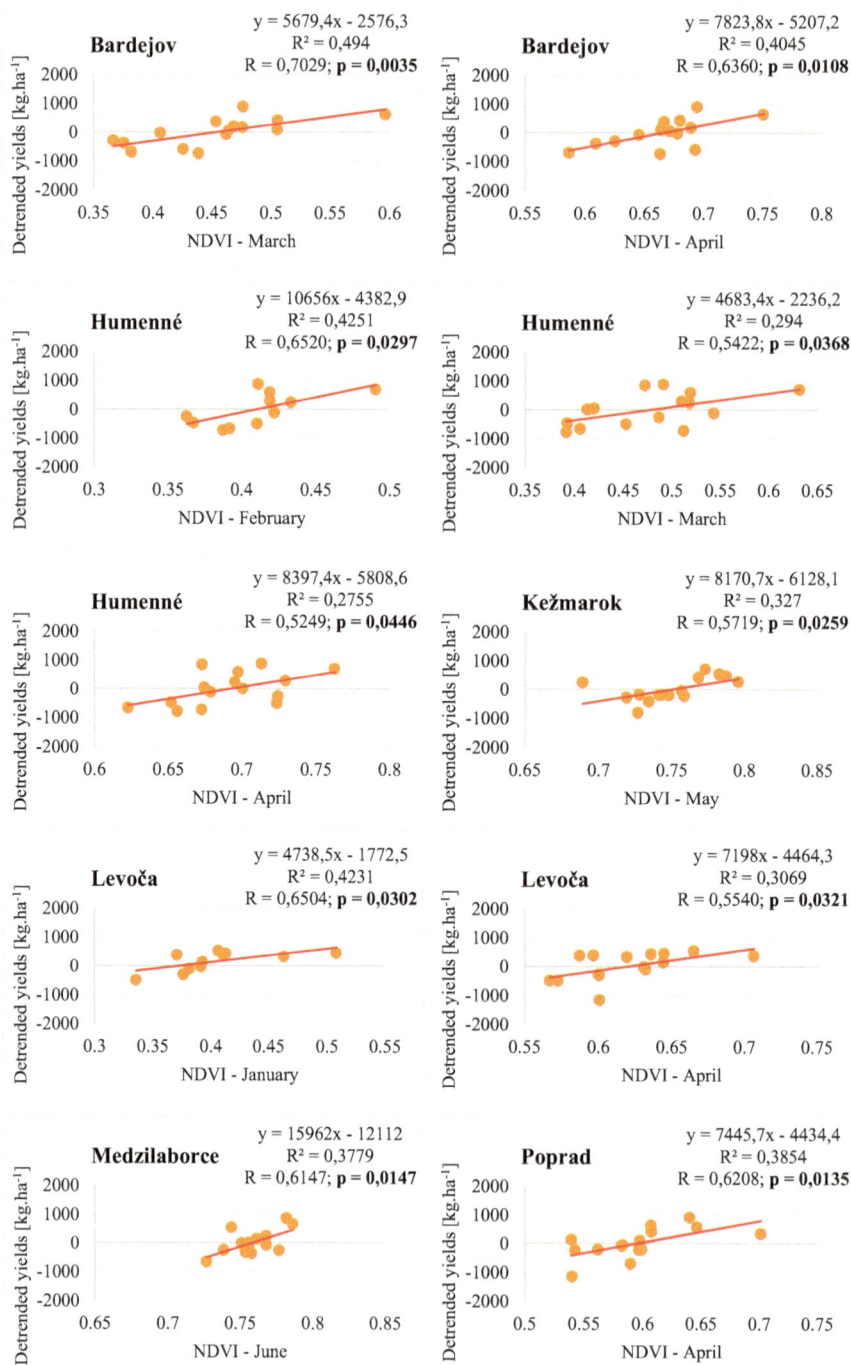

Fig. 4.18 Statistically significant correlations between detrended winter wheat yields in kg.ha⁻¹ and monthly NDVI values in individual districts in Prešov region (2000–2014)

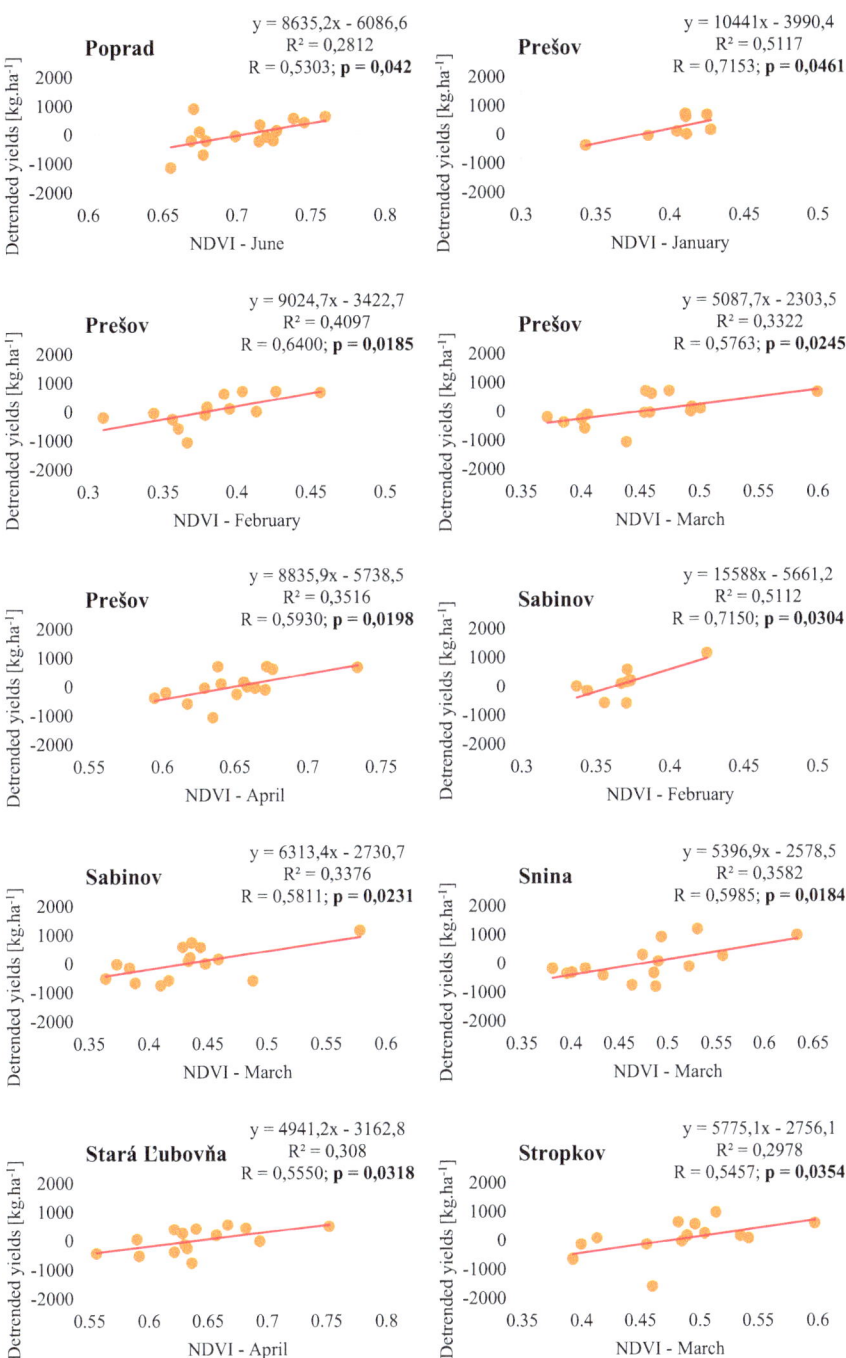

Fig. 4.18 (continued)

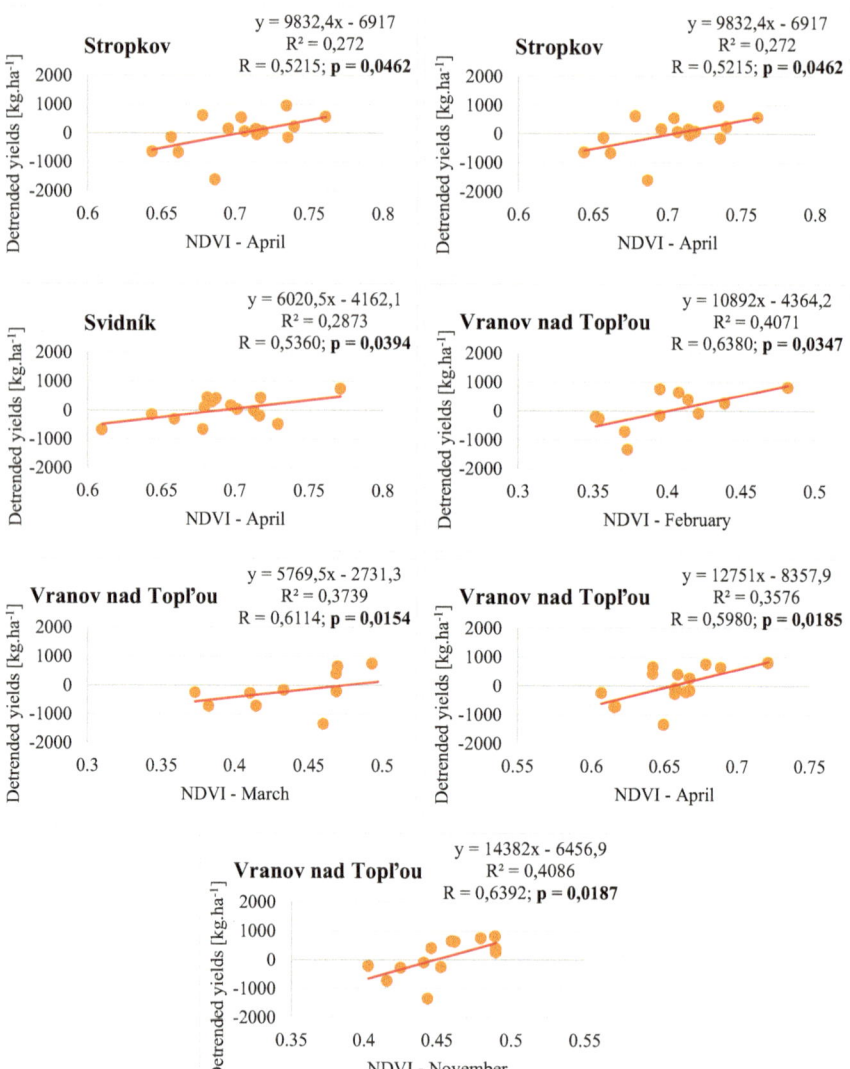

Fig. 4.18 (continued)

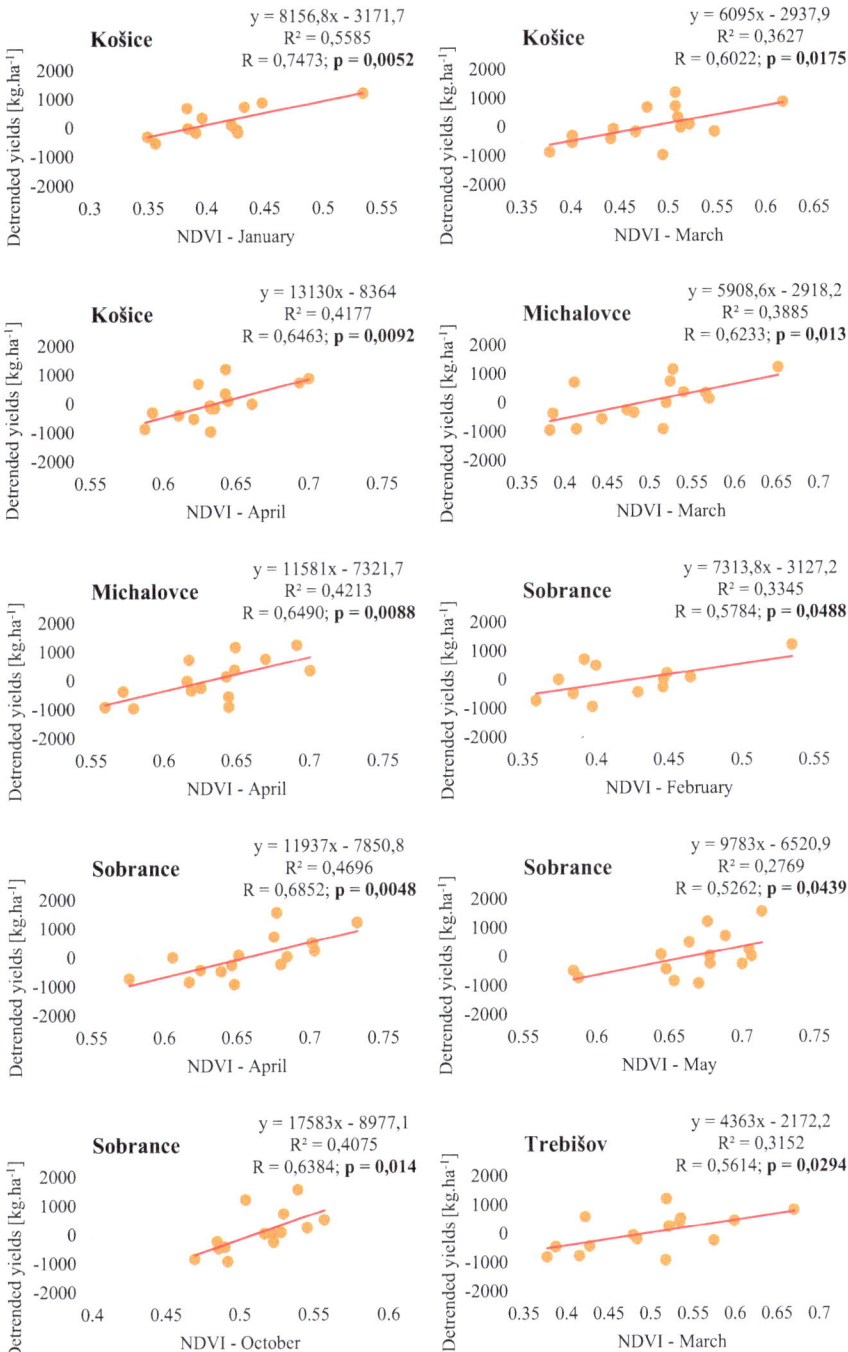

Fig. 4.19 Statistically significant correlations between detrended winter wheat yields in kg.ha^{-1} and monthly NDVI values in individual districts in Košice region (2000–2014)

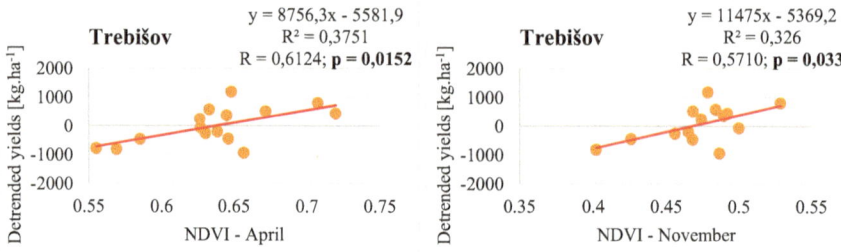

Fig. 4.19 (continued)

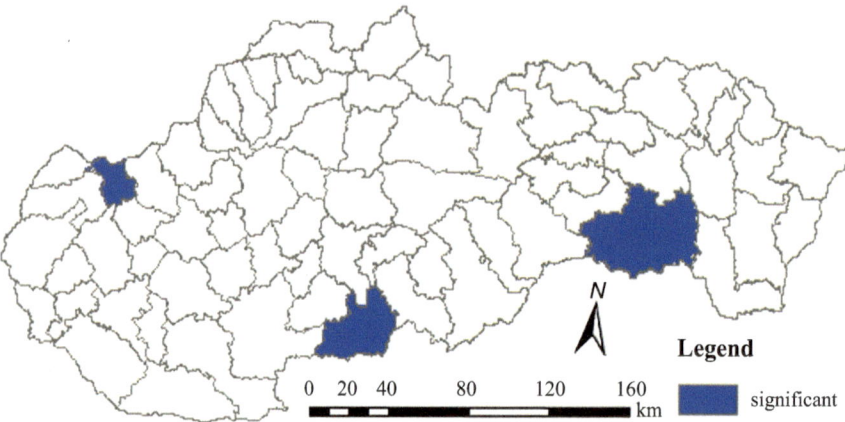

Fig. 4.20 Statistically significant (p ≤ 0.01) correlations between detrended yields of winter wheat [kg.ha⁻¹] and NDVI values on a district level in January during the period 2000–2014

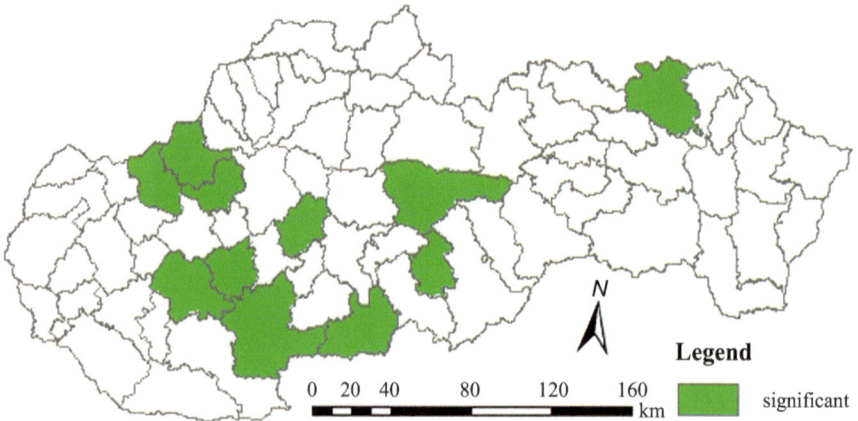

Fig. 4.21 Statistically significant (p ≤ 0.01) correlations between detrended yields of winter wheat [kg.ha⁻¹] and NDVI values on a district level in March during the period 2000–2014

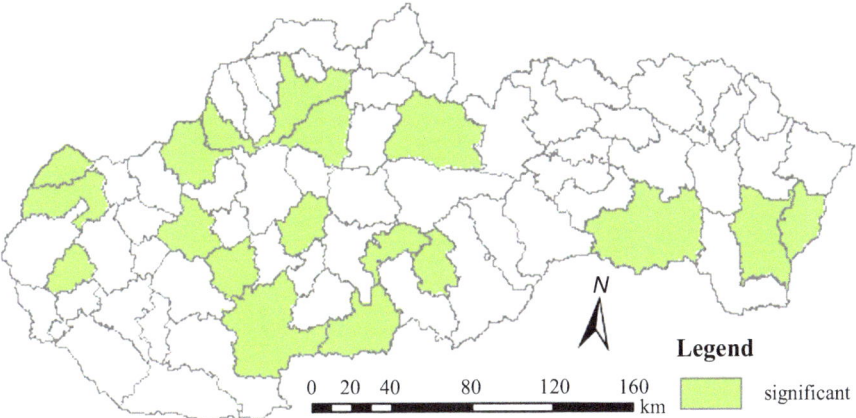

Fig. 4.22 Statistically significant (p ≤ 0.01) correlations between detrended yields of winter wheat [kg.ha^{-1}] and NDVI values on a district level in April during the period 2000–2014

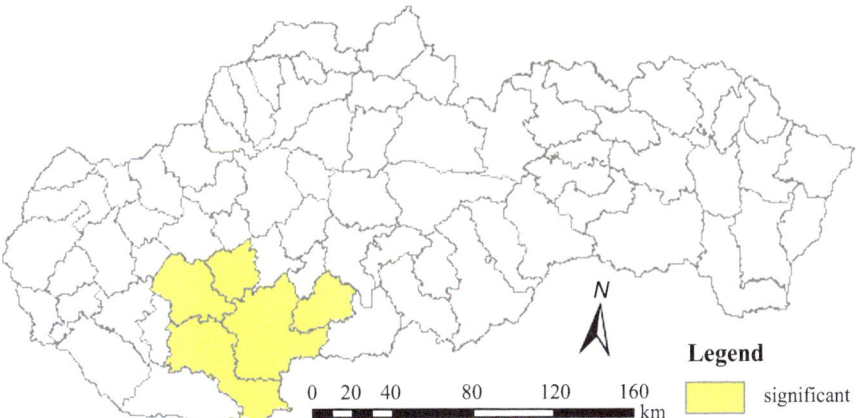

Fig. 4.23 Statistically significant (p ≤ 0.01) correlations between detrended yields of winter wheat [kg.ha^{-1}] and NDVI values on a district level in May during the period 2000–2014

The recommendation for further research is to verify the used index derived from satellite imagery by correlation with other indices of drought. We used PDSI in this work, but it represents a cumulative index. This means that the drought intensity in 1 month depends not only on the weather of the month but also on the intensity of drought in the previous months. Perhaps it is the reason that we evaluated the relationships at the level of statistical significance α ≤ 0.1. An index suitable for verification could be the Z-index, which evaluates the short-term drought on a monthly basis. Consequently, an extremely dry month can be observed during a long-term wet period and vice versa, a wet month can be observed during a long-term period of drought (Karl 1986). It would also be useful to compare the NDVI with other

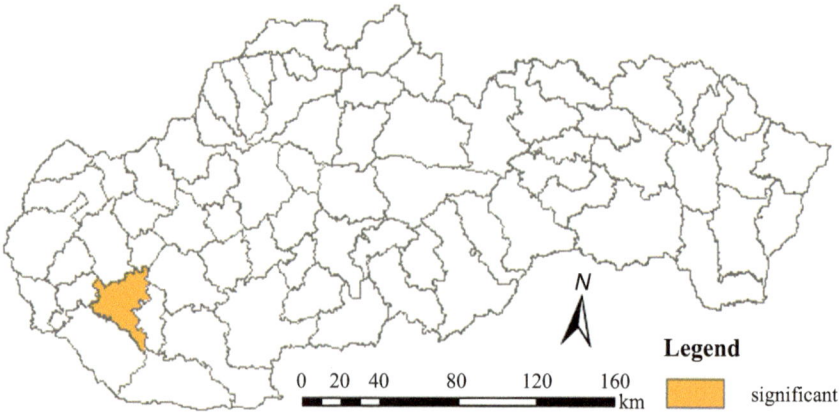

Fig. 4.24 Statistically significant (p ≤ 0.01) correlations between detrended yields of winter wheat [kg.ha⁻¹] and NDVI values on a district level in June during the period 2000–2014

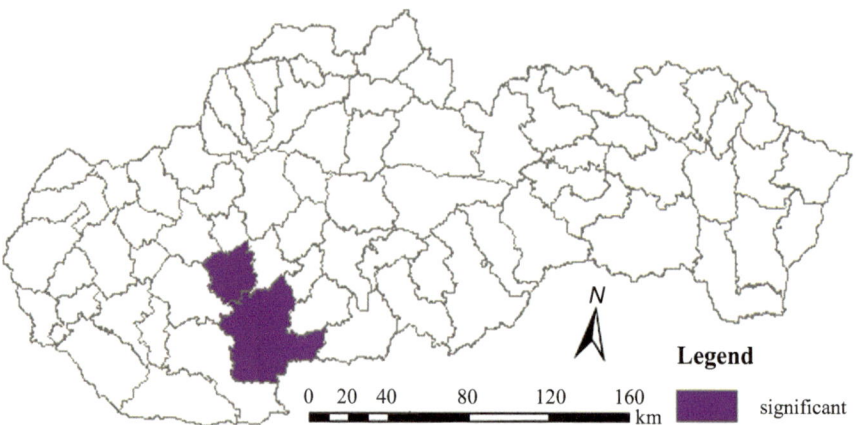

Fig. 4.25 Statistically significant (p ≤ 0.01) correlations between detrended yields of winter wheat [kg.ha⁻¹] and NDVI values on a district level in October during the period 2000–2014

vegetation indices to determine their reaction to drought. The results would allow the determination of the most suitable vegetation index for the assessment of drought in Slovakia.

Another recommendation is to continue the following works focused on the drought assessment by NDVI and to supplement the results with other cultivated field crops. Since MODIS data (MOD13Q1) at 250 m are available since 2000, only a 15-year period was evaluated. Continuation of collection and evaluation this data would provide statistically significant analyses of a longer time series.

The results of the work are a suitable base for irrigation management on arable land in Slovakia. The used methods allow to quantify the lack of moisture in the various regions and months of the growing season of the evaluated field crops.

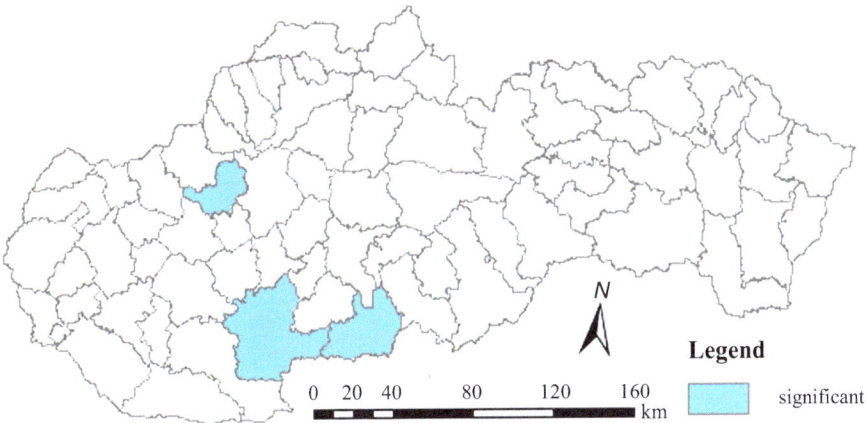

Fig. 4.26 Statistically significant (p ≤ 0.01) correlations between detrended yields of winter wheat [kg.ha^{-1}] and NDVI values on a district level in November during the period 2000–2014

Fig. 4.27 Statistically significant correlations between detrended maize yields in kg.ha^{-1} and monthly NDVI values in individual districts in Bratislava region (2000–2014)

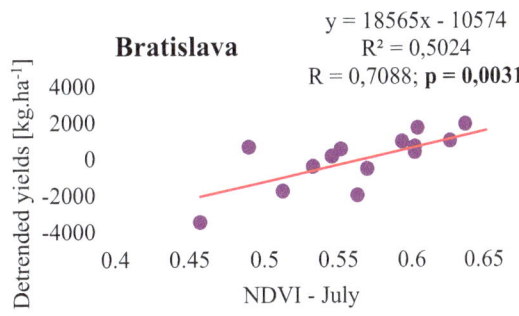

Irrigation is an important factor in crop production, but also in the production of vegetables and fruits. In order to maintain the stability and production of agro-ecosystems in agricultural production areas, water from additional irrigation will be a prerequisite for optimizing the soil water regime. However, even irrigation water needs to be used efficiently and with an emphasis on saving it.

The processed vegetation index values can be used to estimate crop yields. Soil Science and Conservation Research Institute (SSCRI) implements a crop yield estimate based on the methodology proposed by the European Commission Joint Research Centre (JRC Ispra) implemented at the national level. The prediction is allowed by the European Crop Growth Monitoring System (CGMS). For this activity, it is necessary to know the state and development of biomass in a particular area, which is ensured by the NDVI (Klikušovská et al. 2015). Also, the forecast for agricultural commodity prices is also based on the knowledge of weather patterns and their impact on cultivated crops.

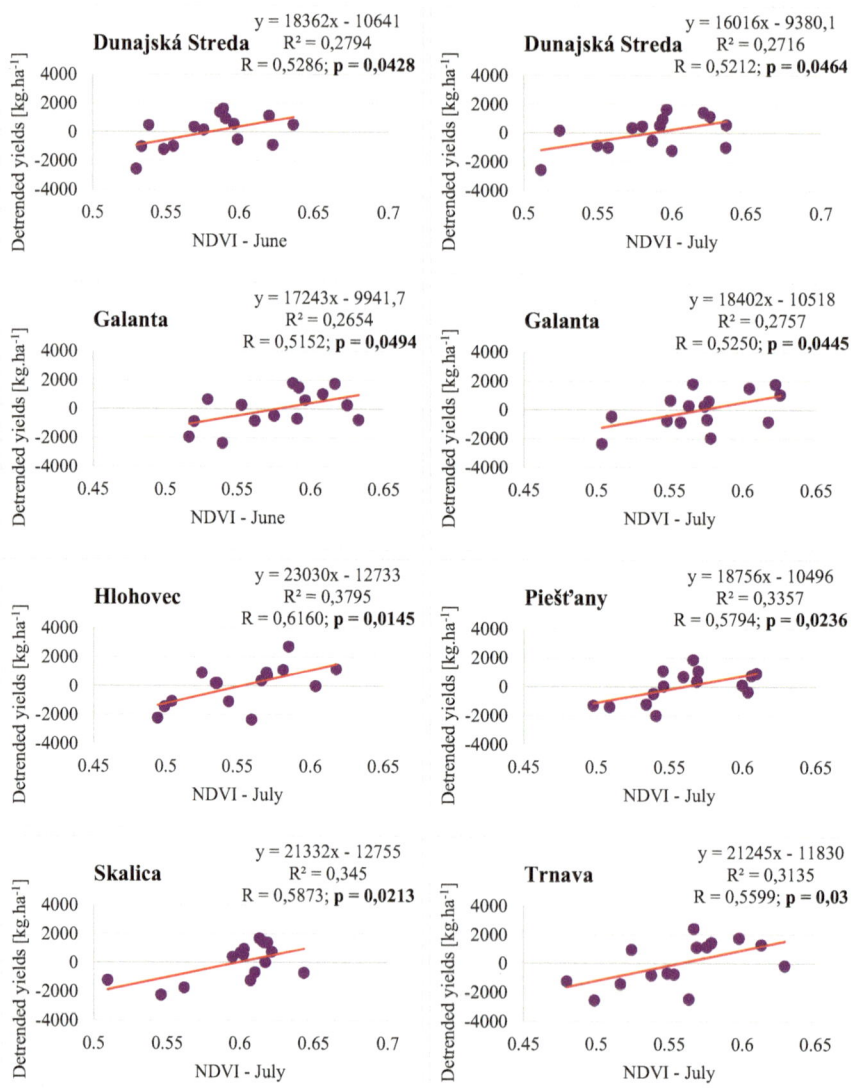

Fig. 4.28 Statistically significant correlations between detrended maize yields in kg.ha⁻¹ and monthly NDVI values in individual districts in Trnava region (2000–2014)

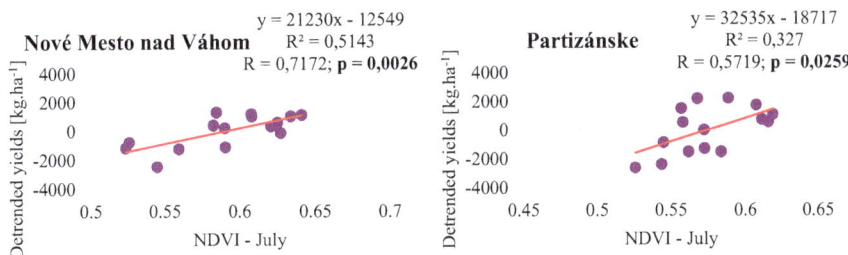

Fig. 4.29 Statistically significant correlations between detrended maize yields in kg.ha⁻¹ and monthly NDVI values in individual districts in Trenčín region (2000–2014)

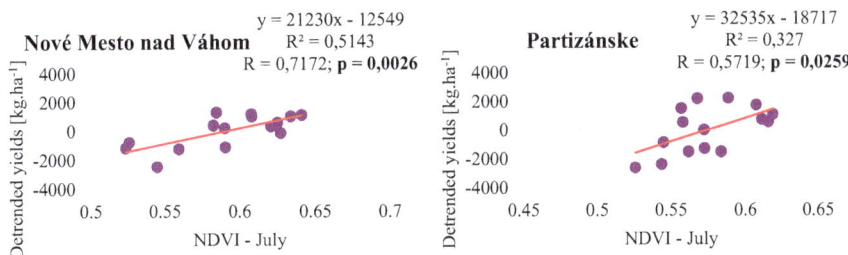

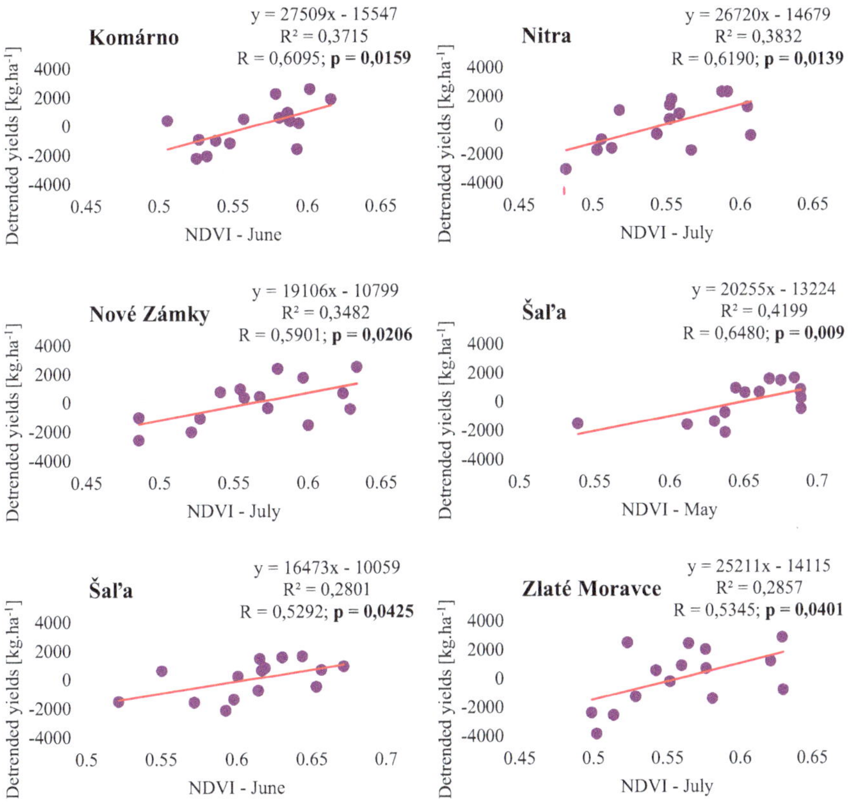

Fig. 4.30 Statistically significant correlations between detrended maize yields in kg.ha⁻¹ and monthly NDVI values in individual districts in Nitra region (2000–2014)

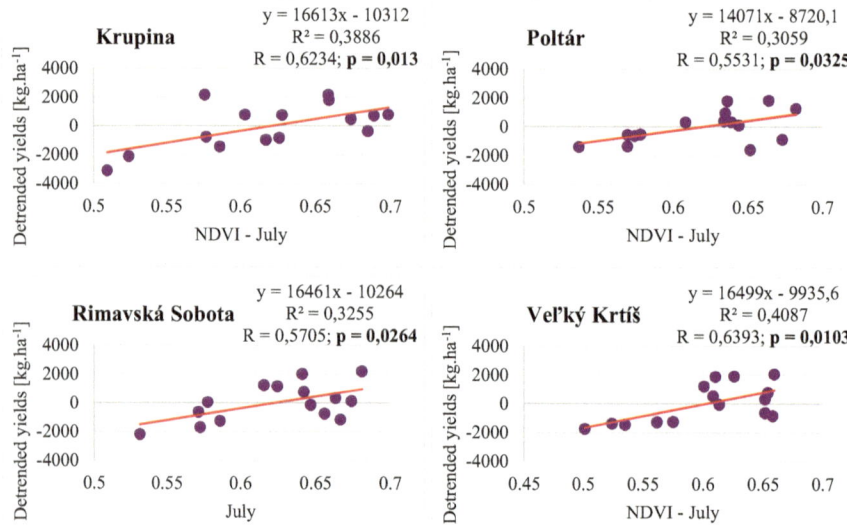

Fig. 4.31 Statistically significant correlations between detrended maize yields in kg.ha⁻¹ and monthly NDVI values in individual districts in Banská Bystrica region (2000–2014)

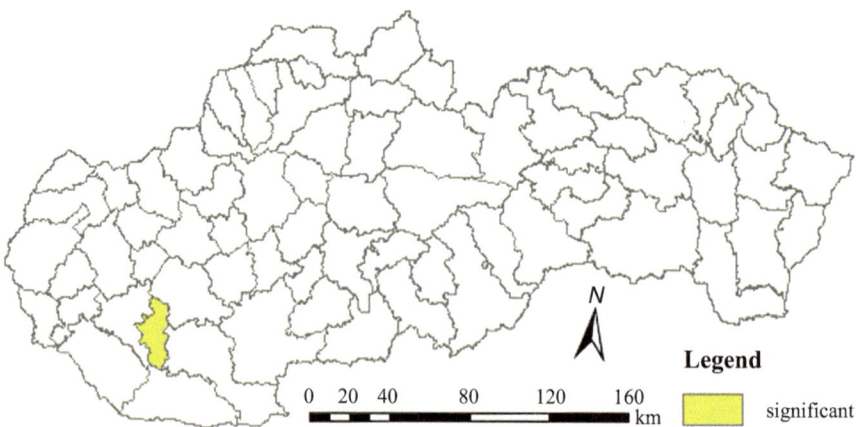

Fig. 4.32 Statistically significant ($p \leq 0.01$) correlations between detrended yields of maize [kg. ha⁻¹] and NDVI values on a district level in May during the period 2000–2014

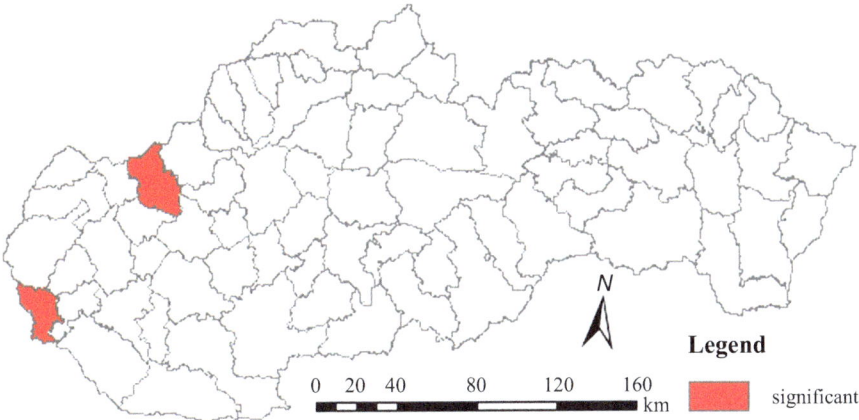

Fig. 4.33 Statistically significant ($p \leq 0.01$) correlations between detrended yields of maize [kg. ha^{-1}] and NDVI values on a district level in July during the period 2000–2014

The results of the work and similar research works serve as a basis for the insurance industry, whose task is to take care of the risks. In changing climatic conditions, demand for crop insurance may increase from damage caused by weather extremes. For valuation models, insurers will require up-to-date scientific knowledge of extreme weather events.

4.1.3 Potential of Applied Methods

Methods of investigation, that were used in the study, make possible to determine critical months and sites in terms of the influence of the moisture deficit on crop yields. Based on the results, appropriate management measures then can be proposed.

Influence of moisture deficiency on spring barley crops was the strongest in May when the drought affects vegetation mainly in Podunajská nížina Valley. As sensitive regions were detected Bratislava, Trnava, and Nitra. The significant correlations between evaluated indicators were observed only in Krupina district from Banská Bystrica region. Based on the assessment at district level we can state that the highest sensitivity of crops was observed in the districts of Pezinok, Senec, Galanta, Senica, Trnava, Komárno, Levice, Nitra, Nové Zámky, Šaľa, Topoľčany, Zlaté Moravce and in Krupina, which was mentioned earlier. Drought did not affect spring barley yields in the region of Trenčín, Prešov, and Košice.

The strongest drought impact on winter wheat yields was found out in April. As sensitive districts were rated Pezinok, Senica, Skalica, Ilava, Trenčín, Levice, Topoľčany, Zlaté Moravce, Liptovský Mikuláš, Martin, Žilina, Detva, Poltár, Veľký

Krtíš, Žiar nad Hronom, Košice, Michalovce and Sobrance. The impact of drought on winter wheat yields did not demonstrate in the Prešov region. However, in Podunajská nížina Valley (the most fertile territory of Slovakia) winter wheat did not react as sensitively as spring barley. The assessment of the impact of drought on the winter wheat by NDVI is limited in winter when the vegetation is under the snow cover.

The maize, which is more tolerant to drought stress, did not respond to drought impact as sensitively as the other two evaluated cereals. The reaction of maize to lack of moisture was observed only in three districts. In Šaľa in May and in Bratislava and Nové Mesto nad Váhom in July. However, when we reduced the level of statistical significance from $\alpha \leq 0.01$ to $\alpha \leq 0.05$, the significant effect of drought in July was confirmed in the districts Bratislava, Dunajská Streda, Galanta, Hlohovec, Piešťany, Skalica, Trnava, Nové Mesto nad Váhom, Partizánske, Nitra, Nové Zámky, Zlaté Moravce, Krupina, Poltár, Rimavská Sobota and Veľký Krtíš. It follows that the most sensitive regions were Trnava and Nitra. Impact of drought on maize yields was not observed in evaluated districts of the region of Prešov. In the Žilina region, these relationships were not evaluated due to insufficient data.

4.2 Results as the Base for Adaptation Strategies in Agriculture in Slovakia

Agricultural crop production is very sensitive to changing climatic conditions, including drought. Consequently, the appropriate adaptation measures to prepare and adapt to climate change must be properly considered and selected. In order to eliminate the impacts of drought on crops, strategies of species composition change, breeding of resistant varieties and introduction of supplemental irrigation were selected.

The change in species composition results from the sensitivity of the evaluated field crops to drought in certain regions of Slovakia. The results show that spring barley was the most sensitive cereal in the production areas of Western Slovakia. In the area of Central and Eastern Slovakia, the drought sensitivity of spring barley has not been confirmed, which implies that these areas are the most suitable for its cultivation. In sensitive areas, it could be replaced by winter wheat, which did not respond very sensitively in Western Slovakia. Similarly, winter wheat sensitivity to drought has not been confirmed in the Prešov region. Concerning the results of the districts in which maize was evaluated, it can be stated that maize has confirmed its drought-resistant characteristics. It is, therefore, a suitable crop in its cultivation areas. Only after the decrease of the significance level from 0.01 to 0.05 did the effect of drought on maize yields in the districts of Nitra and Trnava regions.

Other selected measures are the support of breeding of new varieties that would be resistant to the variability of climatic conditions and the use of supplemental irrigation to ensure a water regime of crop production. This solution is costly since reconstruction and modernization of existing irrigation systems and purchase of additional irrigation equipment are necessary. The use of supplemental irrigation must be carried out efficiently and with an emphasis on saving water resources (Ministry of Environment of the Slovak Republic 2013; Minďaš et al. 2011). Irrigation as an effective adaptation measure for stabilization of crop yields was proposed e.g. by authors Takáč et al. (2011), who evaluated specific crops in simulations with and without irrigation. They state that the highest field crop yields were recorded in conditions under irrigation in the Podunajská nížina Valley in the time series of 1961–1990. While without irrigation the yields of selected crops in this area were the lowest.

The implementation of selected adaptation strategies would make it possible to ensure the economic efficiency of the agricultural sector and the food security of the population.

Chapter 5
Conclusion

Abstract The chapter presents a summary of the achieved results, which consist of the evaluation of drought using PDSI, derivation of NDVI from remote sensing imagery and correlations between PDSI and NDVI and NDVI and selected field crop yields. The strongest drought impact on spring barley yields was observed in May, on winter wheat yields were observed in April. Maize did not react so sensitively to drought, but the strongest relationship was observed in July. Adaptation strategies appropriate for future cultivation in Slovakia were also proposed.

Keywords Results · Drought · Adaptation strategies

In this study the usage of remotely sensed NDVI as a method suitable for an evaluation of drought in agricultural land was presented. It was preceded by evaluation of drought using PDSI, which uses not only meteorological, but also pedological characteristics of a region (Alley 1984).

Subsequently, the correlations between values of NDVI and values of PDSI during the growing season (April–June) in the period 2000–2014 were evaluated by a linear function. Effect of drought (PDSI) on vegetation vigor (NDVI) was manifested mainly in June, July, August, and September. We tested a number levels of significance, but we detected 0.1 as optimal. The necessity of diminution of significance level might be caused due to cumulative effect of PDSI. It means that the intensity of drought in a current month does not depend only on current conditions, but also on recent conditions to a certain extent. The potential of NDVI as a suitable method for drought assessment was confirmed. Increase of PDSI values leads to increase of NDVI values and vice versa. It means that impact of drought on vegetation is reflected by worse vegetation vigor.

Evaluation of the effect of drought on selected field crop yields was investigated by correlation between detrended yields of individual crops and monthly values of NDVI. The most serious effect of drought on spring barley yields was observed in Bratislava, Nitra and Trnava region in May. This was demonstrated by correlation

V. Zuzulová et al., *Agricultural Drought in Slovakia: An Impact Assessment*, SpringerBriefs in Environmental Science, https://doi.org/10.1007/978-3-030-42061-1_5

coefficients ranging from 0.64 to 0.82 and coefficients of determination from 0.41 to 0.67 in individual districts.

The strongest drought impact on winter wheat yields was observed in April in some districts of almost every region in Slovakia, except for Prešov region. Values of correlation coefficients were found out from 0.64 to 0.78 and values of coefficients of determination from 0.41 to 0.61. However, in comparison with spring barley, winter wheat was not so sensitive to drought in the production area of Podunajská nížina Valley. During the growing season this crop uses water supplies in winter seasons.

Maize did not react so sensitively to drought. When the level of significance was decreased from 0.01 to 0.05 (compared to the previous two evaluated cereals), the effect of drought was detected mainly in July in districts of Nitra and Trnava region. The calculated correlation coefficient was ranging from 0.52 to 0.74 and coefficients of determination from 0.27 to 0.51. This method is suitable for evaluation of winter wheat and spring barley (C3 plants). Maize (C4 plant) is adapted for drier conditions.

Adaptation strategies appropriate for future cultivation in Slovakia and stabilization of evaluated crop production in conditions of changing climate could be supplemental irrigations, breeding drought-resistant varieties and change of species composition in cultivated field crops.

References

Alley WM (1984) The Palmer Drought Severity Index: limitations and assumptions. J Clim Appl Meteorol 23(7):1100–1109. https://doi.org/10.1175/1520-0450(1984)023<1100:TPDSIL>2.0 .CO;2

Backlund P, Janetos A, Schimel DS, Hatfield J, Ryan M, Archer S, Lettenmaier D (2008) Executive summary. In: Backlund P, Janetos A, Schimel D (eds) The effects of climate change on agriculture, land resources, water resources, and biodiversity in the United States. Climate Change Science Program, Washington DC, pp 1–10. https://www.fs.fed.us/rm/pubs_other/rmrs_2008_backlund_p003.pdf. Accessed 23 Aug 2019

Baret F, Guyot G (1991) Potentials and limits of vegetation indices for LAI and APAR assessment. Remote Sens Environ 35(2–3):161–173. https://doi.org/10.1016/0034-4257(91)90009-U

Baret F, Guyot G, Major D (1989) TSAVI: a vegetation index which minimizes soil brightness effects on LAI and APAR estimation. 12th Canadian symposium on remote sensing geoscience and remote sensing symposium, 3, pp 1355–1358. https://doi.org/10.1109/IGARSS.1989.576128

Berhan G, Hill S, Tadesse T, Atnafu S (2011) Using satellite images for drought monitoring: a knowledge discovery approach. J Strateg Innov Sustain 7(1):135–153. https://repository.upenn.edu/oid_papers/264. Accessed 23 Aug 2019

Biswal A, Jeyaram A, Mukherjee S, Kumar U (2013) Analysis of temporal and spatial changes in the vegetation density of Similipal Biosphere Reserve in Odisha (India) using multitemporal satellite imagery. Int J Ecol 2013: 6p. https://doi.org/10.1155/2013/368419

Brandýsová V, Bucha T (2012) Vplyv prízemnej vegetácie a podrastu na priebeh fenologickej krivky bukových porastov odvodenej z údajov MODIS (Effect of understory vegetation and undergrowth on course of phenological curve of beech forests derived from MODIS). Lesnícky časopis (For J) 58(4):231–242. http://www.nlcsk.sk/files/3587.pdf. Accessed 23 Aug 2019

Brázdil R, Trnka M, Dobrovolný P, Chromá K, Hlavinka P, Žalud Z (2009) Variability of droughts in the Czech Republic, 1881–2006. Theor Appl Climatol 97(3–4):297–315

Brown JF, Wardlow BD, Tadesse T, Hayes MJ, Reed BC (2008) The Vegetation Drought Response Index (VegDRI): a new integrated approach for monitoring drought stress in vegetation. GISci Remote Sens 45(1):16–46. https://doi.org/10.2747/1548-1603.45.1.16

Bucha T, Priwitzer T, Koreň M (2011) Modelovane fenologického vývoja lesných porastov pomocou vegetačného indexu NDVI odvodeného zo satelitných snímok MODIS (Modelling phenological development of forest stands using vegetation index NDVI derived from satellite scenes MODIS). Lesnícky časopis (For J) 57(3):187–196. http://www.nlcsk.sk/files/2668.pdf. Accessed 23 Aug 2019

© The Author(s), under exclusive license to Springer Nature Switzerland AG 2020
V. Zuzulová et al., *Agricultural Drought in Slovakia: An Impact Assessment*,
SpringerBriefs in Environmental Science,
https://doi.org/10.1007/978-3-030-42061-1

Büntgen U, Brázdil R, Frank D, Esper J (2010) Three centuries of Slovakian drought dynamics. Clim Dyn 35:315–329. https://doi.org/10.1007/s00382-009-0563-2

Dobrovolný P (1998) Dálkový průzkum Země, Digitální zpracování obrazu (Remote sensing, digital image processing). Masarykova univerzita, Brno, 210p

Dubrovsky M, Svoboda MD, Trnka M, Hayes MJ, Wilhite DA, Zalud Z, Hlavinka P (2009) Application of relative drought indices in assessing climate-change impacts on drought conditions in Czechia. Theor Appl Climatol 96(1–2):155–171

Dufour JM (2011) Coefficients of determination, 12p. http://www2.cirano.qc.ca/~dufourj/Web_Site/ResE/Dufour_1983_R2_W.pdf. Accessed 23 Aug 2019

Dunkel Z (2009) Brief surveying and discussing of drought indices used in agricultural meteorology. Időjárás 113(1–2):23–37

Fendeková M, Gauster T, Labudová L, Vrablíková D, Danáčová Z, Fendek M, Pekárová P (2018) Analysing 21st century meteorological and hydrological drought events in Slovakia. J Hydrol Hydromech 66(4):393–403

Gilabert MA, Gonzáles-Piqueras J, García-Haro FJ, Meliá J (1998) Designing a generalized soil-adjusted vegetation index (GESAVI). In: Engman ET (ed) Proceedings of SPIE 3499, remote sensing for agriculture, ecosystems, and hydrology, 11 December 1998, Barcelona, Spain, pp 396–404. https://doi.org/10.1117/12.332774

Gilabert MA, Gonzáles-Piqueras J, García-Haro FJ, Meliá J (2002) A generalized soil-adjusted vegetation index. Remote Sens Environ 82(2–3):303–310. https://doi.org/10.1016/S0034-4257(02)00048-2

Gobron N (2008) Leaf area index (LAI). In: Sessa R, Dolman H (eds) Terrestrial essential climate variables for climate change assessment, mitigation and adaptation [GTOS 52], pp 32–33. http://www.fao.org/3/i0197e/i0197e15.pdf. Accessed 23 Aug 2019

Hlavinka P, Trnka M, Semerádová D, Dubrovský M, Žalud Z, Možný M (2009) Effect of drought on yield variability of key crops in Czech Republic. Agric For Meteorol 149(3–4):431–442

Hrvoľ J, Horecká V, Škvarenina J, Střelcová K, Škvareninová J (2009) Long-term results of evaporation rate in xerothermic Oak altitudinal vegetation stage in southern Slovakia. Biologia 64(3):605–609

Huang J, Wang X, Li X, Tian H, Pan Z (2013) Remotely sensed rice yield prediction using multi-temporal NDVI data derived from NOAA's AVHRR. PLoS One 8(8):e70816. https://doi.org/10.1371/journal.pone.0070816

Huete AR (1988) A Soil-Adjusted Vegetation Index (SAVI). Remote Sens Environ 25(3):295–309. https://doi.org/10.1016/0034-4257(88)90106-X

Huete A, Didan K, Miura T, Rodriguez EP, Gao X, Ferreira LG (2002) Overview of the radiometric and biophysical performance of the MODIS vegetation indices. Remote Sens Environ 83(1–2):195–213. https://doi.org/10.1016/S0034-4257(02)00096-2

IPCC (2014) Climate change 2014: synthesis report. In: Core Writing Team, Pachauri RK, Meyer LA (eds) Contribution of working groups I, II and III to the fifth assessment report of the intergovernmental panel on climate change. IPCC, Geneva, 151p

Jackson RD, Huete AR (1991) Interpreting vegetation indices. Prev Vet Med 11(3–4):185–200. https://doi.org/10.1016/S0167-5877(05)80004-2

Jiang Z, Huete AR, Didan K, Miura T (2008) Development of a two-band enhanced vegetation index without a blue band. Remote Sens Environ 112(10):3833–3845. https://doi.org/10.1016/j.rse.2008.06.006

Karl TR (1986) The sensitivity of the Palmer Drought Severity Index and Palmer's Z-Index to their calibration coefficients including potential evapotranspiration. J Clim Appl Meteorol 25(1):77–86. https://doi.org/10.1175/1520-0450(1986)025<0077:TSOTPD>2.0.CO;2

Karnieli A, Agam N, Pinker RT, Anderson M, Imhoff ML, Gutman GG, Panov N, Goldberg A (2010) Use of NDVI and land surface temperature for drought assessment: merits and limitations. J Clim 23(3):618–633. https://doi.org/10.1175/2009JCLI2900.1

Kasawani I, Norsaliza U, Mohd Hasmadi I (2010) Analysis of spectral vegetation indices related to soil-line for mapping mangrove forests using satellite imagery. Appl Remote Sens J 1(1):25–31

Klikušovská Z, Kusý D, Sviček M (2015) Odhad úrody a produkcie pšenice letnej formy ozimnej, jačmeňa siateho jarného a kapusty repkovej pravej (Estimation of production of winter wheat, spring barley and rapeseed), Správa k 15.07.2015, Bratislava: Národné poľnohospodárske a potravinárske centrum, Výskumný ústav pôdoznalectva a ochrany pôdy, 23p. http://www.podnemapy.sk/portal/verejnost/akt_poln_sezona/_vystupy/Odhad_15.7.2015.pdf. Accessed 23 Aug 2019

Koch B, Ammer U, Schneider T, Wittmeier H (1990) Spectroradiometer measurements in the laboratory and in the field to analyse the influence of different damage symptoms on the reflection spectra of forest trees. Int J Remote Sens 11(7):1145–1163. https://doi.org/10.1080/01431169008955085

Kurpelová M, Coufal L, Čulík J (1975) Agroklimatické podmienky ČSSR (Agroclimatic conditions CSSR). Bratislava: Príroda, 270p

Labudová L, Turňa M, Nejedlík P (2015) Drought monitoring in Slovakia. In: Šiška B, Nejedlík P, Eliašová M (eds) Towards climatic services. International scientific conference, September. Slovak University of Agriculture, Nitra, pp 15–18

Labudová L, Labuda M, Takáč J (2017) Comparison of SPI and SPEI applicability for drought impact assessment on crop production in the Danubian Lowland and the East Slovakian Lowland. Theor Appl Climatol 128(1–2):491–506

Lapin M, Gera M, Kremler M (2010) Scenáre zmeny teploty a vlhkosti vzduchu na Slovensku a možné dôsledky v mestách (Air temperature and humidity scenarios in Slovakia and their potential impacts in cities). Životné prostredie (Environment) 44(5):227–231

Leblon B (1993) Soil and vegetation optical properties. In: The International Center for Remote Sensing Education (ed) Applications in remote sensing 4. Wiley, New York

Litschmann T, Rožnovský J (2001) Palmerův index závažnosti sucha a jeho aplikace pro lokalitu Žabčice (Application of Palmer Drought Severity Index for Žabčice). In: Rožnovský J, Janouš D (eds) Sucho, hodnocení a predikce: Pracovní seminář. Brno, ČHMÚ

Litschmann T, Klementová E, Rožnovský J (2002) Vyhodnocení period sucha v časových řadách pražského Klementina a Hurbanova pomocí PDSI (Assessment of dry periods in time series from observatory in Prague – Klementinum and Hurbanovo by PDSI). In: Rožnovský J, Litschmann T (eds) XIV. Česko-slovenská bioklimatologická konference, Lednice na Moravě, 2–4 September 2002, pp 280–289

LP DAAC, NASA, U.S. Geological Survey (2016) MOD13Q1: MODIS/Terra Vegetation Indices 16-Day L3 Global 250m Grid SIN V006. https://lpdaac.usgs.gov/dataset_discovery/modis/modis_products_table/mod13q1_v006. Accessed 17 Dec 2016

LP DAAC, U.S. Geological Survey, EROS (2011) MODIS reprojection tool, User's manual, Release 4.1. https://lpdaac.usgs.gov/sites/default/files/public/mrt41_usermanual_032811.pdf. Accessed 1 Aug 2016

Maselli F (2004) Monitoring forest conditions in a protected Mediterranean coastal area by the analysis of multiyear NDVI data. Remote Sens Environ 89(4):423–433. https://doi.org/10.1016/j.rse.2003.10.020

McKee TB, Doesken NJ, Kleist J (1993) The relationship of drought frequency and duration to time scales. In: Eighth conference on applied climatology, 17–22 January 1993. Anaheim, American Meteorological Society, pp 179–184

Meneses-Tovar CL (2009) Analysis of the Normalized Differential Vegetation Index (NDVI) for the detection of degradation of forest cover in Mexico 2008–2009: case studies on measuring and assessing forest degradation. Forest resources assessment working paper 173, 23p. http://www.fao.org/docrep/012/k8593e/k8593e00.pdf. Accessed 23 Aug 2019

Minďaš J, Páleník V, Nejedlík P et al. (2011) Dôsledky klimatickej zmeny a možné adaptačné opatrenia v jednotlivých sektoroch, záverečná správa (Consequences of climate change and possible adaptation measures in individual sectors, final report), Zvolen, 252p. http://www.shmu.sk/File/projekty/Zaverecna%20Sprava%20projektu%20Klim.%20zmena%20a%20Adaptacie%202012.pdf. Accessed 23 Aug 2019

Ministry of Environment of the Slovak Republic (2013) Stratégia adaptácie SR na nepriaznivé dôsledky zmeny klímy, 68p. http://www.shmu.sk/File/ExtraFiles/SHMU_AKTUALITY/files/Strategia_adaptacie_SR_draft.pdf. Accessed 23 Aug 2019

Moghaddam A, Ardekani AS, Karami J (2015) Identifying destructive fire in the fire affected areas using satellite images (Case study: forest fires in Golestan province in 2011). Int J Sci Eng Appl Sci 1(7):519–528. http://ijseas.com/volume1/v1i7/ijseas20150756.pdf. Accessed 23 Aug 2019

Mróz M, Sobieraj A (2004) Comparison of several vegetation indices calculated on the basis of a seasonal SPOT XS time series, and their suitability for land cover and agricultural crop identification. Tech Sci 7:39–66. http://www.uwm.edu.pl/wnt/technicalsc/ts7_2004/4_7_2004.pdf. Accessed 23 Aug 2019

Myneni RB, Hall FG, Sellers PJ, Marshak AL (1995) The interpretation of spectral vegetation indexes. IEEE Trans Geosci Remote Sens 33(2):481–486. https://doi.org/10.1109/36.377948

NASA (2016a) Terra. https://eospso.gsfc.nasa.gov/sites/default/files/mission_handbooks/Terra.pdf. Accessed 23 July 2019

NASA (2016b) Aqua. https://eospso.gsfc.nasa.gov/sites/default/files/mission_handbooks/Aqua.pdf. Accessed 23 July 2019

NASA, EOSDIS (2016) Reverb ECHO. https://reverb.echo.nasa.gov/reverb/#utf8=%E2%9C%93&spatial_map=satellite&spatial_type=rectangle&keywords=mod13q1. Accessed 1 Aug 2016

Nováková M, Klikušovská Z, Skalský R, Sviček M, Mišková M, Čičová T (2010) Národný systém pre odhad úrod a produkcie poľnohospodárskych plodín SK_CGMS (National system for estimating yields and crop production SK_CGMS). Bratislava: VUPOP, 32p

Palmer WC (1965) Meteorologic drought: research paper no. 45. Washington, DC: U.S. Weather Bureau, 58p

Palutikof JP, Goodess CM, Guo X (1994) Climatic change, potential evapotranspiration and moisture availability in the Mediterranean basin. Int J Climatol 14(8):853–869. https://doi.org/10.1002/joc.3370140804

Panda SS, Ames DP, Panigrahi S (2010) Application of vegetation indices for agricultural crop yield prediction using neural network techniques. Remote Sens 2(3):673–696. https://doi.org/10.3390/rs2030673

Pearson RL, Miller LD (1972) Remote mapping of standing crop biomass for estimation of the productivity of the shortgrass prairie, Pawnee National Grasslands, Colorado. In: Proceedings of the 8th international symposium on remote sensing of the environment 2, pp 1355–1379

Peters AJ, Walter-Shea EA, Ji L, Viña A, Hayes M, Svoboda MD (2002) Drought monitoring with NDVI-based standardized vegetation index. Photogramm Eng Remote Sens 68(1):71–75. https://pdfs.semanticscholar.org/778b/cf08400090a59c9f9aef95ac0dd5af0cae14.pdf. Accessed 23 Aug 2019

Porra RJ (2002) The chequered history of the development and use of simultaneous equations for the accurate determination of chlorophylls a and b. Photosynth Res 73(1–3):149–156. https://doi.org/10.1023/A:1020470224740

Portela MM, Zelenákova M, Santos JF, Purcz P, Silva AT, Hlavatá H (2015) Drought analysis in Slovakia: regionalization, frequency analysis and precipitation thresholds. WIT Trans Ecol Environ 197:237–248. https://doi.org/10.2495/RM150211

Potopová V, Štěpánek P, Možný M, Türkott L, Soukup J (2015a) Performance of the standardised precipitation evapotranspiration index at various lags for agricultural drought risk assessment in the Czech Republic. Agric For Meteorol 202:26–38

Potopová V, Zahradníček P, Türkott L, Štěpánek P, Soukup J (2015b) The effects of climate change on variability of the growing seasons in the Elbe River Lowland, Czech Republic. Adv Meteorol 2015:1–16

Qi J, Chehbouni A, Huete AR, Kerr YH, Sorooshian S (1994) A modified Soil Adjusted Vegetation Index. Remote Sens Environ 48(2):119–126. https://doi.org/10.1016/0034-4257(94)90134-1

Richardson AJ, Wiegand CL (1977) Distinguishing vegetation from soil background information. Photogramm Eng Remote Sens 43(12):1541–1552

Rimarčík M (2014) Dvojrozmerná induktívna štatistika – intervalové premenné (Two-dimensional inductive statistics – interval variables). http://rimarcik.com/navigator/interval2.html. Accessed 10 Oct 2016

Rojas O (2013) Operational maize yield model development and validation based on remote sensing and agro-meteorological data in Kenya. Int J Remote Sens 28(17):3775–3793. https://doi.org/10.1080/01431160601075608

Rondeaux G, Steven M, Baret F (1996) Optimization of soil-adjusted vegetation indices. Remote Sens Environ 55(2):95–107. https://doi.org/10.1016/0034-4257(95)00186-7

Rouse JW Jr, Haas RH, Deering DW, Schell JA (1973) Monitoring the vernal advancement and retrogradation (green wave effect) of natural vegetation: progress report RSC 1978-2. http://ntrs.nasa.gov/archive/nasa/casi.ntrs.nasa.gov/19740004927.pdf. Accessed 23 Aug 2019

Šiška B, Takáč J (2009) Drought analyse of agricultural regions as influenced by climatic conditions in the Slovak Republic. Időjárás 113(1–2):135–143

Škvarenina J, Tomlain J, Hrvoľ J, Škvareninová J (2009a) Occurrence of dry and wet periods in altitudinal vegetation stages of West Carpathians in Slovakia: time-series analysis 1951–2005. In: Střelcová K (ed) Bioclimatology and natural hazards. Springer, Dordrecht, pp 97–106

Škvarenina J, Tomlain J, Hrvol J, Škvareninová J, Nejedlík P (2009b) Progress in dryness and wetness parameters in altitudinal vegetation stages of West Carpathians: time-series analysis 1951–2007. Idojárás 113(1–2):47–54

Škvarenina J, Vido J, Minďaš J, Střelcová K, Škvareninová J, Fleischer P, Bošeľa M (2018) Globálne zmeny klímy a lesné ekosystémy (Global climate change and forest ecosystems). Technická Univerzita vo Zvolene, Zvolen

Statistical Office of the Slovak Republic (2019) Search center. www.statistics.sk. Accessed 23 Aug 2019

Steven MD (1998) The sensitivity of the OSAVI vegetation index to observational parameters. Remote Sens Environ 63(1):49–60. https://doi.org/10.1016/S0034-4257(97)00114-4

Střelcová K, Minďáš J, Škvarenina J (2006) Influence of tree transpiration on mass water balance of mixed mountain forests of the West Carpathians. Biologia 61(19):305–310

Sultana SR, Ali A, Ahamd A, Mubeen M, Zia Ul Haq M, Ahmad S, Ercisli S, Jaafar HZE (2014) Normalized difference vegetation index as a tool for wheat yield estimation: a case study from Faisalabad, Pakistan. Sci World J 2014: 8p. doi:https://doi.org/10.1155/2014/725326

Šustek Z, Vido J (2013) Vegetation state and extreme drought as factors determining differentiation and succession of carabid communities in the forests damaged by the windstorm in High Tatra in 2004. Biologia 68(6):1198–1210

Šustek Z, Vido J, Škvareninová J, Škvarenina J, Šurda P (2017) Drought impact on ground beetle assemblages (Coleoptera, Carabidae) in Norway spruce forests with different management after windstorm damage – a case study from Tatra Mts. (Slovakia). J Hydrol Hydromech 65(4):333–342. https://doi.org/10.1515/johh-2017-0048

Takáč J (2013) Assessment of drought in agricultural regions of Slovakia using soil water dynamics simulation. Agriculture 59(2):74–87

Takáč J, Šiška B, Nováková M (2011) Možné dôsledky zmeny klímy na potenciál úrod poľných plodín na južnom Slovensku (Climate change impacts on yield potential of field crops in southern Slovakia). In: Salaš P (ed) Rostliny v podmínkách měnícího se klimatu, Lednice, 20–21 October 2011, Úroda, vědecká příloha, pp 612–622

Tall A (2008) Application of the palmer drought severity index in east Slovakian lowland. Cereal Res Commun 36(1):1195–1198

Tall A, Gomboš M (2011) Aplikácia Palmerovho indexu pre hodnotenie sucha (Application of Palmer Drought Severity Index for evaluation of drought). In: Salaš P (ed) Rostliny v podmínkách měnícího se klimatu. Lednice, 20–21 October 2011, Úroda, vědecká príloha, pp 623–628

Thornthwaite CW (1948) An approach towards a rational classification of climate. Geogr Rev 38(1):55–94. https://doi.org/10.2307/210739

Trnka M, Balek J, Štěpánek P, Zahradníček P, Možný M, Eitzinger J, Žalud Z, Formayer H, Turňa M, Nejedlík P, Semerádová D, Hlavinka P, Brázdil R (2016) Drought trends over part of Central Europe between 1961 and 2014. Clim Res 70:143–160. https://doi.org/10.3354/cr01420

Tucker CJ, Townshend JRG, Goff TE (1985) African land-cover classification using satellite data. Science 227(4685):369–375. https://doi.org/10.1126/science.227.4685.369

U.S. Geological Survey (2015) NDVI, the Foundation for Remote Sensing Phenology. http://phenology.cr.usgs.gov/ndvi_foundation.php. Accessed 23 Aug 2019

Valach J, Vido J, Škvarenina J (2016) Výskyt sucha v regióne Horného Požitavia v období 1966–2013 (Drought occurrence in the Horné Požitavie region over the period 1966–2013). Acta Hydrologica Slovaca 17(1):30–36

Vicente-Serrano SM, Beguería S, López-Moreno JI (2010) A Multiscalar Drought Index sensitive to global warming: the Standardized Precipitation Evapotranspiration Index. J Clim 23(7):1696–1718. https://doi.org/10.1175/2009JCLI2909.1

Vido J, Valach J, Škvarenina J (2014) Zhodnotenie výskytu sucha použitím indexu SPI v regióne Horného Požitavia. In: Rožnovský J, Litschmann T, Středa T, Středová H (eds.) Extrémy obĕhu vody v krajinĕ. Mikulov, 8–9 April 2014, Praha: ČHMÚ, 10p

Vido J, Tadesse T, Šustek Z, Kandrík R, Hanzelová M, Škvarenina J, Škvareninová J, Hayes M (2015) Drought occurrence in central European mountainous region (Tatra National Park, Slovakia) within the period 1961–2010. Adv Meteorol 2015:1–8

Vido J, Nalevanková P, Valach J, Šustek Z, Tadesse T (2019) Drought analyses of the Horné Požitavie region (Slovakia) in the period 1966–2013. Adv Meteorol 2019:1–10

Vilček J, Škvarenina J, Vido J, Nalevanková P, Kandrík R, Škvareninová J (2016) Minimal change of thermal continentality in Slovakia within the period 1961–2013. Earth Syst Dynam 7(3):735–744

Viňa A, Gitelson AA, Nguy-Robertson AL, Peng Y (2011) Comparison of different vegetation indices for the remote assessment of green leaf area index of crops. Remote Sens Environ 115(2):3468–3478. https://doi.org/10.1016/j.rse.2011.08.010

Weier J, Herring D (2000) Measuring vegetation (NDVI & EVI). http://earthobservatory.nasa.gov/Features/MeasuringVegetation/. Accessed 23 Aug 2019

Willstätter R, Stoll A (1913) Untersuchungen über Chlorophyll: Methoden und Ergebnisse. Julius Springer, Berlin

Žalud Z, Trnka M, Kapler P, Semerádová D, Dubrovský M (2006) Sucho – problém současnosti i budoucnosti (Drought – present and future meteorological hazard). Kvasný průmysl 52(7–8):203–234. http://kvasnyprumysl.cz/pdfs/kpr/2006/07/04.pdf. Accessed 23 Aug 2019

Zeleňáková M, Vido J, Portela M, Purcz P, Blištán P, Hlavatá H, Hluštík P (2017) Precipitation trends over Slovakia in the period 1981–2013. Water 9(12):922

Zhang H, Chen H, Zhou G (2012) The model of wheat yield forecast based on MODIS-NDVI: a case study of Xinxiang. ISPRS Ann Photogramm Remote Sens Spat Inf Sci 1(7):25–28. https://doi.org/10.5194/isprsannals-I-7-25-2012. Accessed 23 Aug 2019

Zuzulová V, Šiška B (2017) Identification of drought in Western Slovakia by Palmer Drought Severity Index (PDSI). Acta Regionalia et Environmentalica 14(1):7–14

Zverko J, Vido J, Škvareninová J, Škvarenina J (2014) Early onset of spring phenological phases in the period 2007–2012 compared to the period 1931–1960 as a potential bioindicator of environmental changes in The National Nature Reserve Boky (Slovakia). In: International conference on "Mendel and bioclimatology". Mendel University, Brno, pp 469–476